Circular Economy

Modern Finance, Management Innovation and Economic Growth Set

coordinated by
Faten Ben Bouheni

Volume 3

Circular Economy

From Waste Reduction to Value Creation

Edited by

Karen Delchet-Cochet

ISTE

WILEY

First published 2020 in Great Britain and the United States by ISTE Ltd and John Wiley & Sons, Inc.

ISTE Ltd
27-37 St George's Road
London SW19 4EU
UK

www.iste.co.uk

John Wiley & Sons, Inc.
111 River Street
Hoboken, NJ 07030
USA

www.wiley.com

MIX
Paper from responsible sources
FSC® C013604

Library of Congress Control Number: 2020935510

British Library Cataloguing-in-Publication Data
A CIP record for this book is available from the British Library
ISBN 978-1-78630-573-2

Contents

Foreword

With the arrival of new words and new concepts, our understanding of what they carry requires a significant period of time to stabilize, first of all in the framework of the first warnings to reach public opinion, through the field of politics in charge of operational implementation.

This is the case with the concept of “sustainable development”. In France, and probably in French-speaking countries, “sustainable development” has come to be commonly summed up as the only ecological requirement, forgetting the other two pillars of sustainable development – social and economic requirements.

So what should we think of the term “circular economy”, which semantically refers to reference points that have no known link with each other? Of course, with these two words, “economy” and “circular”, what do we do with them? The answer lies in Nature, which in essence has not forgotten the limits of the planet: a blue orange has been circling the Sun for billions of years; of these billions of years and the billions that follow, it will not grow a gram heavier. Like this sailboat launched non-stop into the ocean of the Universe with what it carried, such is our Earth. Waste and destroy the resources available on board and that is the end of humanity’s destiny. Fortunately, as human beings, we are extraordinarily fortunate to be able to understand the limits of our own existence, our place in the galaxy and therefore the need to preserve our resources.

However, recycling waste into a new raw material cannot be the only answer. In a few decades, the Earth will be home to 10–12 billion people, including 5–7 billion who will be able to possess as much as we have, as they will have been lucky enough to possess things in our so-called developed countries. In this scenario, two or three Earths would not be enough. And we only have one!

One Earth with abundant but limited resources.

A rapidly growing population multiplied by 10 in 150 years.

A model of development known as linear economy, based on voracious levying, wasteful processing and wasteful consumption.

The only parameter that can be changed in order to close this parenthesis on a century and a half of linear economy, is our development model. This is the very essence of the circular economy, which shares the Earth's resources among all its inhabitants today, tomorrow and in the centuries to come so that everyone can have the means to live without suffering shortages, conflicts and wars over raw materials.

This is the challenge, to rethink our development to achieve this balance between growth for the common good and preservation of resources. Our current model, based solely on the creation of economic wealth, artificial in the sense of the physical reality of the planet, must incorporate a new dimension in order to continue to create values, certainly values that are economic, as well as social and societal, while at the same time reducing the amount of resources we take from it: quite a challenge.

It is the aim of this book, which, in a cross-disciplinary research of transdisciplinarity, provides answers for the implementation via policies of the new foundations of our development model. Based on the experience of pioneering companies, on work in sociology and education, on the shared normative framework and on the indispensable return to territories as spaces for sustainable development, this book enables everyone to grasp this paradigm shift, this revolution, by approaching it by author's contribution, by thematic chapter or as a whole.

I would like to commend this work, a remarkable work of reference today, to be read as a matter of urgency. It will not be out of date for years to come, even if it is clear that new ideas, new concepts and innovative models will emerge, because the foundations are real foundations for those who want to understand the circular economy in relation to the many challenges posed by the limits of our planet and the framework of our unique way of thinking, which is blocked by a development model, the societal rules that structure our common life and the dynamics of human growth.

Supervised by Karen Delchet-Cochet, the experts, many of whom I know and whom I thank for what they have brought to my understanding of the issues, give us the bricks that we now need to put together to build a society for ever more shared common goods without wasting our planet's resources.

In my contributions, I am pleased to recall that the most important thing in a circular economy approach is the coffee machine, a place that brings together those who should not meet and who will share their expertise and co-construct the project together. This book is a paper coffee machine. Annotate this book, alter it, share it, enrich it! Make it a book of circular economy, bringing in more and more knowledge; collective knowledge.

I reiterate my admiration to all the contributors and my special thanks go to Karen Delchet-Cochet, the initiator and editor of this book.

François-Michel LAMBERT
Member of French Parliament
President of the *Institut national de l'économie circulaire*

Introduction

The circular economy is an exciting topic, both for the present and the future since it is nothing more than a system overhaul, to paraphrase the Ellen MacArthur Foundation[1]. At a time of climate change, the collapse of biodiversity, and increasing inequality, "the Earth is reacting to human actions", explained Bruno Latour in his inaugural lesson to Sciences Po[2,3] students in 2019. We are facing a second "Galilean revolution". For Lacy and Rutqvist, "transitioning to the circular economy may be the biggest revolution and opportunity in our global economy"[4]. However it is also necessary to understand what the circular economy means, which we believe is already too often reduced to a system adaptation.

Beyond the interest and topicality of the subject, the need for a collective work originates from a triple observation. The first observation is the absolute need to engage experts from different disciplines in a dialogue to collectively address the many challenges we face both in terms of resources and equity. Most of the time, we remain confined to our fields of research or expertise. The topic of the circular economy is, by nature, transdisciplinary. In other words, "it goes beyond the silos between disciplines"[5]. It, therefore, requires perspectives and proposals between them, i.e., interdisciplinary. The second observation is the need for academic research to inform operational debate and implementation. The French legislative calendar for 2019 was conducive to discussions in terms of the circular economy.

---

1 https://www.ellenmacarthurfoundation.org/fr/economie-circulaire/concept.

2 https://www.sciencespo.fr/actualites/actualit%C3%A9s/%E2%80%9Cce-n%E2%80%99est-plus-une-question-d%E2%80%99%C3%A9cologie-mais-de-civilisation%E2%80%9D/4306.

3 Sciences Po is a prestigious French higher education institution. It has notable alumni, including previous heads of state and government including seven of the eight past French Presidents.

4 Lacy, P. and Rutqvist, J. (2015). *Waste to Wealth*. Palgrave Macmillan, London.

5 https://www.larousse.fr/dictionnaires/francais/transdisciplinaire/79087.

Indeed, the "*Anti gaspillage pour une économie circulaire*" ("Anti waste for a circular economy") bill has launched a debate in which we are participating here. We wish to stress the need to consider the circular economy as a resource economy rather than a waste economy, a path that one now seems to be taking. The final observation, as a corollary to the first two points, is the importance of a collective reflection with multiple perspectives.

This book, therefore, includes reflections by environmentalists, urban planners, and specialists in management sciences, including the perspective of researchers, business consultants, local actors, and institutions.

They base their comments on theories, observations and business examples. They shed light on what the circular economy covers and allow us to *understand its framework and issues in Part 1*. Thus, if the circular economy is not an entirely new concept, it seems, as Franck Aggeri says in Chapter 1 of this book, necessary to "invent a new model", a "mode 2" circular economy. He emphasizes the creation of local value and the importance of a new industrial policy.

Christian Brodhag stresses the need to return to the notion of resources and the environmental context (Chapter 2). He highlights the importance of a lifecycle approach and of an environmental, social, and economic assessment to innovate without negative impacts.

This paradigm shift towards a circular economy necessarily leads to an evolution of the labor market and human resources management, the main trends of which David Moriez studies (Chapter 3). New jobs and new skills are emerging.

This evolution, therefore, implies that we must reflect on the role of education in the ecological and circular transition, both in terms of initial and continuing training, as Dominique Bonet Fernandez points out in Chapter 4. Student mobilizations and higher education commitments are a strong testimony to this responsibility.

Finally, the work of Karine Fabre and Alexis Pokrovsky concludes Part 1 by focusing on proximity as a lever for a circular economy (Chapter 5). Local and regional authorities are now equipped with new skills to be part of a circular economy. They illustrate their point with examples of public practices and policies between responsible purchasing and waste reduction at the source.

Part 2 presents some examples of tools and approaches. Anne de Bethencourt looks back at the history and philosophy of Cradle to Cradle (Chapter 6). This pioneering approach proposes to eliminate the notion of waste and create positive loops. Several sectors, such as textiles, carpets, and cleaning products, have already deployed it.

Industrial and territorial ecology is closely linked to the circular economy, as Valérie Fernandes points out (Chapter 7). After having reviewed the common points and differences between these two concepts, she wonders about the modes of governance that would ensure the sustainability of industrial and territorial ecology approaches, by giving many examples.

The sharing economy can also be an interesting approach from a circular economy perspective. According to Catherine Lejealle, the sharing economy rightly contributes to the debate around the circular economy (Chapter 8). She highlights the role of consumers in a circular economy and the potential leverage of the sharing economy to play this role.

Finally, Anne Benady, Melodie Merenda and Mérylle Aubrun argue that the circular economy, a subject under construction, required a normative framework, as explained by the Afnor authors' group. They review the genesis of the XP X30-901 standard: "Circular economy – Circular economy project management system – Requirements and guidelines" and present us with its content and challenges, particularly at the international level (Chapter 9).

Lastly, Part 3 of this book focuses on a few sectors of activity with high environmental stakes. Vincent Augiseau provides a comprehensive overview of the issues, definitions, policies, and projects of the circular economy in construction (Chapter 10). He stresses the variability of the scope of the circular economy, insisting in particular on the insufficient consideration of the territory, and the need to articulate circular economy and spatial planning.

Yvon Pesqueux returns to the notion of waste and what it reflects in terms of society in Chapter 11. He discusses the ambiguity of waste and its lifecycle, and questions the boundaries of this stigma in our societies.

Fashion is a highly topical sector in terms of circular economy which Bénédicte Bourcier-Béquaert, Karen Delchet-Cochet, and Valérie Fernandes are interested in (Chapter 12). Subject to extended producer responsibility (EPR) regulations, this sector has substantial environmental and social impacts. They carry out a critical analysis of the circularity practices of major fast-fashion retailers from a logistical, marketing, and strategic perspective.

Another sector subject to EPR is packaging. It is questioned with regard to the circular economy by François Cabaret in Chapter 13. Returning to the different functions of packaging, he discusses the real need in terms of the service provided and suggests some possible actions.

Finally, Karen Delchet-Cochet and Coralie Damay examine, from both an environmentalist and a market researcher point of view, a sector that is somehow special as it affects children: the toy sector. While environmental and social impacts imply rethinking the lifecycle of toys, the very question of the utility of the toy deserves to be asked (Chapter 14).

These authors' comments are complementary. They define and analyze the circular economy from different angles, but all stress the need to not consider the circular economy from a restrictive angle. The circular economy should not be limited to waste management. They call for a paradigm shift in their academic and/or professional wishes and commitments, including the question of resources. May this book make it possible to establish a dialogue between all stakeholders and to collectively innovate to rethink our relationship with the world.

Dr. Karen DELCHET-COCHET
ISC Paris Business School
France

PART 1

The Circular Economy, Between Framework and Stakes

1

The Circular Economy: Historical Perspective and Contemporary Issues

1.1. Introduction

Over the past 10 years, the circular economy has experienced spectacular interest from public authorities, the media, economic and social actors and, more generally, the general public. This enthusiasm is based on a vulgate illustrated by a short promotional film of one and a half minutes that can be viewed online on the European Union's website[1]. What does this film tell us? Each individual (here a European) consumes an increasing amount of raw materials (14 tons per inhabitant of Europe) and generates an increasing volume of waste (5 tons per inhabitant of Europe). These products, resources and waste could, on the other hand, be repaired, reused or recycled. This is the principle of circular economy. It is therefore necessary, as is explained, to move away from the linear economic model that has gradually become structured since the industrial revolution and has eventually become dominant. This linear economy is based on the idea of a world of infinite resources, which can be taken without limits and transformed into products, which in turn will be consumed and then disposed of in landfills. Conversely, the circular economy model aims to close the flows of materials and energy circulating in the economy. Several strategies could be used to achieve this: reducing the quantities of materials and energy actually used to produce goods, extending their lifespan through sharing, repair and reuse, or recycling the materials they contain at the end of their life, according to an endless cycle.

Chapter written by Franck AGGERI.

1 www.europarl.europa.eu/news/fr/headlines/priorities/ecirculaire-dechets/20151201STO05603/ economie-circulaire-definition-importance-etbenefices.

As this short film highlights, circular economy is nowadays presented in the form of a utopian narrative that draws a new economic model and an ecologically sustainable society. This is where a crucial explanatory factor for the current success of the notion lies: the common view of the circular economy does not threaten blood and tears, or the specter of degrowth as a condition for saving the planet; conversely, it suggests that another mode of growth, more virtuous, and based on the principles of economy and systematic reuse of resources, is possible.

However, the expert and historian are left in doubt: is the circular model really new? As for the linear model, how far back does it really go? And if it is not so old, what did the economic model that preceded it look like?

In short, to better understand the contemporary challenges of the circular economy, a historical perspective is essential. We will first briefly recall the origin of the concept and its placement on the public agenda. We will then return to the history of so-called circular practices. This detour will allow us to highlight that the circular economy historically preceded the linear economy model, which only emerged at the end of the 19th Century and only became definitively established after World War II. However, this historical model of the circular economy, which we will call mode 1, has disappeared for reasons that we will explain and that remain valid. This historical detour will allow us to highlight the conditions of the new circular economy model, which we will call mode 2, to meet contemporary requirements, in terms of pollution traceability, ecological, economic and social sustainability. Finally, we will conclude with an analysis of the main obstacles to a circular transition.

1.2. From the origins of the reflections on the circular economy to its inclusion on the public agenda

The idea of circularity of material and energy flows is not new. It can be found as early as 1966 in Kenneth Boulding's book, which advocates that men must find their place in an ecological cyclical system capable of a continuous reproduction of any material form. The very notion of circular economy first appeared in a book on environmental economics in 1989 (Pearce and Turner 1989).

However, it was not until the early 2010s that the concept became truly popular internationally, with the publication of several reports that would help to popularize the concept among decision-makers and the general public.

The notion was quickly seized by the public authorities. As early as 2016, the European Union adopted a circular economy package that has since been translated into an action plan for the circular economy. In France, the publication of an ADEME report in 2013, followed by the first conference on the circular economy in

2014, at the initiative of the *Institut national de l'économie circulaire* (INEC), which brought together more than 2,000 people in Paris, were the first markers of the future enthusiasm. In 2016, the Energy Transition Act made it one of the pillars of its action. Finally, in April 2018, the roadmap on the circular economy, which prefigures the draft law currently being debated in the National Assembly on the fight against waste and the circular economy, which echoes the circular economy package adopted by the European Parliament in December 2017, constitutes the last step in placing this concept on the public agenda.

1.2.1. *The reasons for the enthusiasm*

How can we explain the enthusiasm for this concept?

The first reason is the historical conditions under which the circular economy project is formulated. The end of the 2000s was a time of "crystallization", to use Hannah Arendt's (1988) formula. In Arendt's analysis, crystallization refers to a time when heterogeneous and disconnected elements are suddenly linked together. Three concomitant events create an environment receptive to the idea of a circular economy: the first element was the boom in commodity prices, which quadrupled between 2000 and 2010 and reminded economic and political actors of their economic dependence on natural resources; the second element was the Chinese embargo on rare-earth elements, which are used in all high-tech applications and which is causing panic among economic and political actors; and the third element was the continued degradation of environmental indicators, which underlines the urgency of the ecological crisis.

Taking advantage of this favorable context, the Ellen MacArthur Foundation (EMF), a new player in the circular economy, created in 2010, partnered with McKinsey to write a report in 2013 that made a big impact. This report proposes a pedagogical problematization of the circular economy, taking up the arguments of established approaches (industrial ecology, ecodesign, Cradle to Cradle, etc.) and integrating them into a coherent system associated with evocative visual schemes. But more importantly, the report highlights the potential of circular economy: not only could this new paradigm reduce environmental impacts by reducing the consumption of natural resources, and encourage reuse and recycling, but also new business models for repair, reuse or recycling could generate thousands of billions of dollars (EMF 13). Reducing the environmental footprint without sacrificing economic growth, by decoupling the latter's consumption from resource and energy consumption, is the great promise that the concept covers and that immediately appeals to economic and political actors.

The second reason for the success of the concept is thus the pedagogical dimension of the proposed approach and the promise of a possible reconciliation of economic and environmental objectives.

The promoters of circular economy, such as the EMF, ADEME or INEC, have not only built the storytelling, putting linear economy and circular economy under tension. They have also sought to produce striking diagrams to visualize the concept of a circular economy. The most well-known educational scheme is that proposed by the Ellen MacArthur Foundation, also known as the "butterfly", which describes different circular economy strategies based on loops of increasing compactness. To avoid landfilling, the most circular solution is to extend the life of products through maintenance and repair. Next come strategies to give the product a second life, such as reuse (for the same use), repurposing (for another application) and remanufacturing (renewing the product through remanufacturing). Functional economy, i.e. the intensification of the use of a product through the sale of associated services (e.g. rental), is another strategy. When none of these strategies are possible, recycling and finally energy recovery remain. From an economic point of view, the most solid strategies are those that retain the most economic value. However, they can cannibalize the sale of new products and therefore imply a change in business model. From an environmental point of view, the most compact loops are, in principle, the ones with the least impact. But this also depends on technological progress. Imagine a product whose new generation is much more environmentally efficient than previous ones. It may then be more interesting to encourage the replacement of old products by new ones, following the example of old vehicles, which public authorities are seeking to remove from the market through scrapping premiums. To rigorously assess the best environmental strategy, methods such as lifecycle assessment can be used. This diagram can also serve as a guide to an eco-design approach, where each company defines criteria and specifications for each strategy to improve maintainability, reparability, reusability and recyclability. These different strategies predate the notion of a circular economy. The interest of the proposed approach is to integrate them into a coherent conceptual framework.

Finally, and thirdly, these promoters have developed a series of reports, models and tools (indicators and standards) to serve as a guide for collective action. In France and abroad, actors such as the EMF, ADEME, AFNOR, *Entreprises pour l'Environnement* or the World Business Council on Sustainable Development (WBCSD) are now actively involved in the production of indicators and benchmarks for the circular economy, taking into account the specific characteristics of each sector[2].

1.2.2. *Circular economy as a new rational utopia*

The combination of these three elements (mobilizing narrative, visual diagrams and tools/models) precisely constitutes the three dimensions of what can be called, following Jean-Louis Metzger (2001), a rational utopia.

According to this author, a rational utopia is constructed around three registers: it takes the form of a mobilizing narrative that articulates both a critique of the existing (e.g. linear economics) and a description of an ideal (e.g. circular economy). It is conceived as a set of powerful images aimed at imbuing collective beliefs (e.g. the "butterfly" scheme); it is structured around tools and models that guide collective action (e.g. circularity indicators, norms and management tools for the circular economy). A rational utopia thus corresponds to the problematized narrative of an ideal society, which is therefore based on images that strike the imagination, and on rational elements (reasoning, modeling, calculations) that are supposed to anchor it in the realizable domain. Rational utopias thus combine the mobilizing properties of utopia with the reassuring properties of reason. In doing so, it is a matter of building collective promises that can aggregate and mobilize heterogeneous actors.

Jean-Louis Metzger suggests that the production of rational utopias is a characteristic of our modern societies since the Enlightenment, at the end of the 18th Century, and are a driving force for social change. Far from being governed solely by rational calculation and interests, our societies need the constant production of new utopias to move forward and transform themselves. The circulation of these rational utopias is all the easier because they are works without authors: their production is collective, anonymous. They appear to the public as neutral constructions, open to multiple interpretations and depoliticized, i.e. free of references to ideologically marked currents of thought or to particular authors, which can be the subject of a wide variety of possible appropriations. This is one of the major differences, it seems to me, between the circular economy and related concepts, more precise as well as more marked, such as industrial ecology or Cradle to Cradle, which are more associated with scientific currents, communities of

2 AFNOR, standard XP-X30-901 on circular economy. Available at: https://normalisation.afnor.org/thematiques/economie-circulaire/; WBCSD (2018). Circular economy metrics landscape Available at: https://docs.wbcsd.org/2018/06/Circular_Metrics-Landscape_analysis.pdf; EPE (2018). Les indicateurs de l'économie circulaire pour les entreprises. Available at: http://www.epe-asso.org/les-indicateurs-de-leconomie-circulaire-pour-les-entreprises-octobre-2018/.

practice and identified authors. But it is this construction, which some consider too ecumenical, that provokes criticism from those who support strong sustainability[3].

1.3. Is the circular economy really a new practice?

Do the characteristics of the circular economy, that we have just briefly described, refer to really new practices? A historical detour is required. The linear economy model became dominant with the rise of consumer society and mass production, i.e. after World War II.

1.3.1. *Circular economy, mode 1*

Forms of circular economy have always existed, as historical research shows. Repair, recycling, reuse and repurposing were common practices until the middle of the 20th Century. The term waste, explains the philosopher François Dagognet (1997), was only introduced in France in the 15th Century. At that time, everything was reused or left to natural degradation.

> With the industrialization of the paper mill and the creation of the paper machine at the end of the 18th Century in England, the price of rags reached new heights, thus enhancing the value of the rag-making business. A sector was born. Similarly, bones had multiple outlets: animal charcoal to bleach beet sugar, ammonia salts, bone tallow, glue, etc. The sludge was used for agriculture.

Nothing was lost, everything was recovered and valued, explains Helen Micheaux in a recent thesis and book (Micheaux, 2017, 2019). Historian Sabine Barles (2005) points out that, until about 1870, there was a spontaneous circulation of matter between the city, industry and agriculture.

Three main changes explain the gradual disappearance of this mode 1 circular economy: the first evolution was the industrial revolution, which introduced three disruptions: the production of cheap energy thanks to coal and steam; the reduction of production and extraction costs of natural resources thanks to mechanization and mass production; and the reduction of costs and delays thanks to the development of mechanized transport. These concomitant transformations then made the materials and products resulting from recovery less economically interesting.

A second change disqualified these products and materials: the development of hygienics from the Pasteurian revolution at the end of the 19th Century. With the

3 Bourg, D. (2019). Available at: https://usbeketrica.com/article/dominique-bourg-l-economie-circulaire-ne-constitue-pas-une-demarche-de-progres.

discovery of microbes, hygienics banished the circulation of waste as one of the main causes of epidemics. Eugène Poubelle, then Prefect of the Seine, became famous by the 1884 decree in which he required landlords to provide tenants with containers for household waste, garbage cans, or, in French, "*poubelles*". This was the beginning of the invention of waste containment solutions that led to the development of landfill as the dominant waste treatment solution in the 20th Century.

The third change, from 1930 onwards, is the development of what Baudrillard (1970) called the "consumer society", i.e. the emergence of consumption not as a response to basic needs but as a means of differentiation, where modern people now live through the objects they consume, leading to a society of superabundance where objects eventually possess them. This consumer society could not have developed without the massive intervention of new professions (marketing, design, communication, advertising) and new techniques that have contributed to creating a desire to always own material objects and to constantly renew them.

1.3.2. *Circular economy, mode 2*

The current challenge of the circular economy is obviously not to return to the previous model of the circular economy but to invent a new one, where the requirements of traceability, hygiene, lower environmental impact and quality are respected.

Hygiene and health issues have not disappeared today, quite the contrary, as shown by the obsession of European public authorities to trace, for example, through the REACH directive, potentially toxic substances found in chemicals and consumer products in general. Brominated flame-retardant plastics are a very good example. These plastics, which are widely used in electrical and electronic products, have an important function: they aim to prevent the ignition of devices with hot batteries. But they have a major disadvantage: they contain heavy metals (chromium, bromine, etc.), which are potentially harmful to health. As a result, their recovery is prohibited and they must be disposed of in landfills. However, in current processing centers, automatic plastic sorting does not guarantee 100% efficiency and some brominated plastic residues can be mixed with other plastics for recycling. The controversy grew when traces of brominated plastics were discovered in children's toys made from recycled plastic in China, where environmental norms are less stringent than in Europe. This example underlines the fact that hygiene issues remain completely topical and currently limit the potential of recovery and recycling and are likely to undermine consumer confidence if doubts remain.

Whether for repair, reuse or recycling, the contemporary circular economy therefore depends on the development of a quality economy where strict compliance with specifications and traceability standards must be accompanied by an increase in the skills of the players in these new industrial and business ecosystems.

1.4. Obstacles to the circular transition to mode 2

If public policies are needed, it is obviously because there is no reason for this transition to take place naturally. As everyone has observed, consumer society is now deeply rooted in our behavior. Planned obsolescence is a common practice of manufacturers who are accelerating the rate of product renewal while seeking to control, like Apple, their reparability.

As for the potential value contained in products and waste, it is certainly proven (precious metals, scrap metal, aluminum, plastics, etc.), but it must also be excluded from illegal trafficking and exports. Experts point out that illegal waste trafficking is the fourth largest source of income for organized crime after drugs, prostitution and illegal gambling. Traffickers have a decisive advantage over legal channels: they do not bear the administrative costs, tax levies or clean-up costs that others incur. These traffic flows are also a source of diffuse pollution since they only recover interesting parts or materials, leaving the polluted parts to be abandoned.

The fragility of some sectors is another obstacle. The French government thus aims to recycle 100% of plastics by 2025. A laudable objective, since only 28% of plastics are currently recovered in Europe (including 14% recycled) compared to 40% that end up in landfills and 32% in the wild, according to a recent study by the EMF (2016). However, its development is not solely due to exogenous variables, such as the price of virgin raw materials (oil for plastic), on which the price of recycled materials is indexed. The difficulty is also due to the lack of structure of industrial sectors, which threatens their sustainability. Indeed, the industrial fabric is still weak in this field. Except for scrap metal, paper, aluminum and precious metals, which are now mature sectors, the other sectors are mainly composed of SMEs operating in niche markets, with limited skills and resources to access quality industrial opportunities. For their part, the major waste manufacturers are only beginning to develop recycling activities, anticipating that their traditional landfill and incineration activities will decrease in the face of pressure to reduce these opportunities.

Concerning repair or reuse, the development of these activities presupposes the emergence of a whole ecosystem of actors covering the entire territory and likely to provide the expected services. As for functional economy, which consists, for the producer who retains ownership of the products, of selling services associated with these products, its development today remains very limited despite promises to reduce environmental impacts by intensifying the use of the products while creating economic value through the sale of services. The emblematic examples are few, and concern primarily industrial customers (B to B) through examples such as photocopying (Xerox), lighting (Signify, for example, Philips lighting) or haulage (Michelin fleet solution) rather than end consumers (B to C).

We see that the obstacles to the transition to a circular economy are many and varied. They lie both in changing consumer behavior and in transforming producers' strategies, in combating trafficking, in structuring recycling, repair or reuse chains and in strengthening requirements for eco-organizations to support the development of these chains.

1.4.1. *Local value creation*

The stakes are nevertheless real because such a transition brings territorialized economic value creation and contains important sources of employment. Developing local reuse and repair, setting up short collection and recycling chains, all this can contribute to the reindustrialization of areas that are currently in difficulty.

In terms of job creation, ADEME estimates that recycling is 20 times more employment-intensive than landfilling. Repair and reuse are also employment-intensive service activities. Local authorities and public authorities alike have clearly understood the challenge of developing territorial strategies to capture and develop these valuable deposits and activities. In May 2019, for example, the French government's roadmap on the circular economy mentioned a target of creating 500,000 jobs by 2025 associated with the circular economy.

To encourage the transition to a circular economy, both French and European public authorities are now trying to bring together a whole series of incentives, information, regulation and support measures for innovation: this is the objective of this roadmap.

1.4.2. *The need for a new industrial policy*

However, it should be noted that beyond incentives and some flagship measures (development of eco-modulation, reparability indices, target, etc.) to achieve ambitious objectives, policies in Europe are significantly lacking a long-term vision. In particular, ambitious industrial policies are lacking. The notion of industrial policy, it must be said, has bad press in Europe and is associated, in France, in the collective unconscious with the centralized policies of the 1960s and 1980s, with the TGV, nuclear power or calculation plan; in other words, with planning policies considered outdated in an era of globalization and intensive innovation.

However, an industrial policy for the circular economy would be necessary to organize and support, in the long term, a series of initiatives that are currently disorganized and fragile, and to overcome economic downturns and market volatility. As we have seen in the case of recycled plastic, the development of new sectors requires a training effort, the production of skills, the emergence of technical and expertise centers, the production of standards and rules, and the stimulation of research and innovation that is lacking.

It is not a question of returning to the old ways but of inventing new forms of public action, territorialized and more participatory, which involve the main actors concerned. The most advanced countries in this area, such as Sweden, are much more proactive than we are. They have renewed their public and industrial policies, in consultation with all stakeholders, by undertaking a complete overhaul of their taxation and public interventions to make their country the leader in climate and circular transition. Similarly, China has revived planning for a more resource-efficient transition. Why would that be a crazy idea? Contemporary financialized capitalism, where the future is negligible in quantity through the mechanism of discounting and short-term financial return, has shown its limits in integrating the challenges of an ecological transition, circular or otherwise. It is a question of inventing other forms of action. There is still time.

1.5. References

Arendt, H. (1988). *Condition de l'homme moderne*. Plon, Paris.

Barles, S. (2005). *L'invention des déchets urbains: France, 1790–1970*. Eyrolles, Paris.

Baudrillard, J. (1970). *La société de consommation*. Denoël, Paris.

Boulding, K.E. (1966). The economics of the coming spaceship earth. In *Environmental Quality in a Growing Economy*, Jarrett, H. (ed.). Resources for the Future/Johns Hopkins University Press, Baltimore, pp. 3–14.

Dagognet, F. (1997). *Des détritus, des déchets, de l'abject. Une philosophie écologique*. Les Empêcheurs de penser en rond, Paris.

Ellen MacArthur Foundation (EMF) (2013). Towards the circular economy volume 1: Economic and business rationale for an accelerated transition. Report managed by McKinsey & Company.

Ellen MacArthur Foundation (EMF) (2016). The new plastics economy: Rethinking the future of plastics & catalysing action. Report.

Metzger, J.L. (2001). Management réformateur et utopie rationnelle. *Cahiers internationaux de sociologie*, (2), pp. 233–259.

Micheaux, H. (2017). Le retour du commun au cœur de l'action collective : le cas de la Responsabilité Élargie du Producteur comme processus de responsabilisation et de co-régulation. PhD thesis, École des Mines, Paris.

Micheaux, H. (2019). *Responsabiliser pour transformer : des déchets aux mines urbaines*. Presses de l'École des Mines, Paris.

Pearce, D.W. and Turner, R.K. (1989). *Economics of Natural Resources and the Environment*. John Hopkins University Press, Baltimore.

2

The Circular Economy and Lifecycle

Before giving a substantive definition of the circular economy, based on principles and the scope of its application, it is necessary to note its contingent aspect to the political and scientific processes of which it is part. Concepts have a life, not to say an ecology; they are born, they confront other concepts that they will modify or supplant. During their life, they are redefined by restricting or extending their field, they spread and decline.

From a scientific point of view, they originate and/or are reinterpreted in different disciplinary fields, taking on different definitions. Broad and polysemous concepts such as the circular economy, industrial ecology or sustainable development will be reduced to interpretations as limited as those of the models to which they refer.

On the political level, the legislative process also takes place in silos. The circular economy law discussed in France at the end of 2019 focuses on recycling. It is based on the principle of *extended producer responsibility* (EPR), which implies that economic actors (manufacturers, distributors, importers), who place products on the market that generate waste, take responsibility for all or part of the management of this waste. This approach focuses the circular economy on downstream and waste management.

The international framework places the circular economy in sustainable consumption and production, one of the 17 Sustainable Development Goals on the 2015–2030 agenda.

Chapter written by Christian BRODHAG.

Finally, ISO standardization has its own agenda, disconnected from the European or UN political agenda, which is integrated into the existing standard system, dominated by management standards (see Figure 2.3).

We will organize the chapter into several stages. The first stage will lead us to question the problem of resources to which the circular economy responds. The second on its articulation with other ecological problems, the third on methods to address the circular economy and the fourth on ongoing political and normative processes. The first two deal with substance (especially matter and energy) and the last two with procedures and tools.

2.1. Resources

Access to resources is an essential condition for development and well-being. Their limits and the conditions under which they are used have led to the concept of sustainable development. The Brundtland report considered

> but ultimate limits there are, and sustainability requires that long before these are reached, the world must ensure equitable access to the constrained resource and reorient technological efforts to relieve the pressure. (WCED 1987)

The International Resource Panel[1] considers that maintaining historical trends would lead to an increase in global material use by 110% by 2060, compared to 2015 levels, from 90 to 190 billion tons in 2060, and that resource use would thus increase from 11.9 tons to 18.5 tons per capita (Oberle *et al.* 2019). This development is not sustainable, as the extraction and production of these resources places a considerable burden on the environment.

On the other hand, decoupling resource use from economic growth and well-being could generate an additional $2 trillion to the global economy. It is in this perspective that the circular economy is embedded by promoting value retention and reducing environmental impacts, while reducing costs and creating economic opportunities. By terminating the open lifecycles of *extraction* → *production* → *consumption* → *waste* by closing the loops, the circular economy contributes to this decoupling in a practical way.

1 The International Resource Panel was established by the United Nations Environment Programme (UNEP) in 2007. It is composed of scientists, qualified in the field of resource management, and its mission is to provide independent scientific assessments for the sustainable use of natural resources and, in particular, their environmental impacts throughout their lifecycle: www.resourcepanel.org.

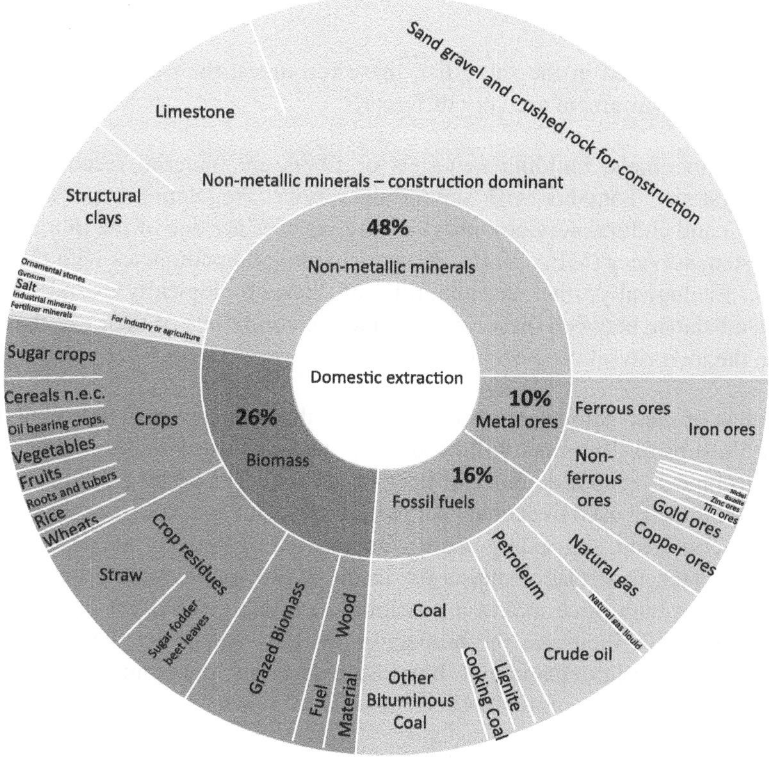

Figure 2.1. *Global domestic mining in 2017, by material group[2]. For a color version of this figure, see iste.co.uk/delchet/circular.zip*

The resources are of very different natures (Figure 2.1). Half of them are non-metallic minerals used mainly in construction. A quarter consists of biomass, and the rest is divided between fossil fuels (16%) and metal ores (10%). The measurement of these resources in tons gives a quantitative view of this pressure, to which qualitative considerations must be added.

Water is not included in this list, whereas with a global consumption of 4000 km^3/year, it is also decisive both for the production of biomass, which consumes 70% of this water, and for the industry (20%).

Since the Earth is a closed system with respect to matter, non-renewable resources are necessarily present in limited stocks. As the Earth is an open system for energy, biomass resources that use solar energy are renewable, but limited in flow.

2 http://www.materialflows.net/visualisation-centre/raw-material-profiles/.

2.2. Environmental context

Although included in the same list, these resources, their availability and their links with the environment are very different.

Food, some fuels, building materials or fibers are material resources derived from ecosystems. Together with supporting services, regulating services (climate, water, etc.) and cultural services, provisioning services are one of the four categories of ecosystem services (MEA 2005). Biomass exploitation competes with these other services. Ecologically, the sustainability of these biologically based resources requires a balance between these services and the conditions of their use that do not threaten the integrity of ecosystem functioning.

The rest of the resources are non-renewable; they concern materials contained in the Earth's subsoil. Reserves will be used to designate recoverable volumes under current technical, economic and political conditions. The reserve/production (R/P) ratio is traditionally used to account for these limits. But as it is expressed in years, it introduces confusion, suggesting that it would be a time limit for the depletion of the resource. In fact, this relationship is the result of a dynamic largely determined by market price: a high price makes it possible to explore and exploit more expensive resources and thus increases the numerator. At the same time, the price increase leads to a decrease in demand and therefore reduces the denominator. For decades, the oil reserve/production ratio has been close to 40 years.

Fossil energy resources (coal, oil, gas, etc.) are largely determined by the price of oil. The decline in resources becoming more scarce was illustrated by the image of Hubbert's peak: the growth in production would be followed by a decrease. This approach places the debate on the date of this reversal. This imminent decline is doubly challenged by new resources and environmental limits. New unconventional resources, such as shale oil and gas, have added 10% to oil reserves and one third to gas reserves, respectively. But from an environmental point of view, these fuels release CO_2, the main greenhouse gas, into the atmosphere. To limit climate change, it is not the reduction of their use that should be targeted, but their elimination. The availability of carbonaceous fossil resources (coal, oil, gas, etc.) is much higher than the atmosphere is able to support. We will not know Hubbert's peak on resources, because the scarcity of the oil resource is not the limiting factor. The limit is not economic but ecological, and therefore political.

For metals, these natural reserves, present in the lithosphere, are competing with materials present in the "*technosphere*" that can be exploited through recycling. Metals such as iron and steel are recycled by 70–90%, manganese by more than 50%, lead or niobium by more than 50% (Graedel *et al.* 2011). On the other hand, rare earths (neodymium, dysprosium, terbium, etc.), which in 2014 represented less than 0.2% of global mining production, are recycled for only 1%. However, these rare earths are essential in the modern economy to produce permanent magnets, phosphors or batteries, present in electrotechnical devices from wind turbines, electric cars to microcomputer hard disks.

But a high recycling rate does not change the situation significantly when uses are increasing. With a recycling rate of 83%, recycled end-of-life steel represents only 16% of the steel produced, 22% comes from manufacturing waste and 62% from ores. In a context of demand growth of approximately 2–3% per year, as is currently the case, recycling only delays the deadline by a few decades at best (Geldron 2017). It is therefore necessary to reduce uses, in addition to targeting recycling percentages.

Materials used in construction and infrastructure represent half of the resources exploited. Some of them are used under problematic environmental conditions, such as the sand needed to make concrete. Two approaches are envisaged: the use of recycled materials, which is directly related to the circular economy, and the use of biosourced materials, which is an additional pressure on ecosystems.

Resource management through the circular economy must therefore be considered environmentally in relation to other global issues such as climate, biodiversity and water. The limits do not only concern the availability of resources and their exploitability, but also the capacity of the biosphere to produce these resources, for biomass, or to absorb pollution, for others.

The exchanges between the economic sphere of life and the industrial sphere, the technosphere, must be carefully evaluated. There are two distinct types of material flows to be considered: biological nutrients, intended to re-enter the biosphere safely, and technical inputs, designed to be recycled while maintaining a high level of quality, without entering the biosphere

An uncontrolled circular economy can be catastrophic. The recycling of animal protein in animal feed led to the mad cow disease crisis. The valorization of organic sludge from wastewater treatment plants in the agricultural amendment can lead to the contamination of soils with heavy metals. Under the guise of a circular economy, waste dilution strategies can prove problematic: industrial residues in the road base, or even, in the future, the recovery of low-level radioactive metals from the dismantling of nuclear power plants. Toxic concentrations considered acceptable at one time are often challenged by further research, which justifies the *a priori* application of the precautionary principle. *The closure and recovery of waste must be subject to an impact assessment.*

Similarly, the extension of the duration of use, another proposal of the circular economy, may also be problematic. This property, valued industrially and economically, can be an environmental disaster. Some very stable molecules, widely used for this purpose, will contaminate the biosphere: PCB endocrine disruptors that accumulate in the food chain, or plastics that contaminate the entire biosphere, until they form a *seventh continent*.

Biomimicry, which draws its inspiration from the forms, materials, properties, processes and functions of living organisms to design products and objects, is also not a systematic guarantee of compatibility with the environment. Finally, the dematerialization model, the functional economy (FE) that aims to substitute the rental of a service for the possession of an object, is also not a guarantee of environmental control, if the rate of renewal is based on the obsolescence of this object, such as the cell phone.

All the concepts used in the circular economy must be carefully evaluated from an environmental point of view, both locally and globally.

2.2.1. *The global question*

The article published in *Nature* in 2009 under the title "A safe operating space for humanity", specified global boundaries that must not be transgressed in order to prevent unacceptable environmental change (Rockström *et al.* 2009). This study was updated in 2015 (Steffen *et al.* 2015). It answers the question of the limitations formulated by the Brundtland report. Quantitative limit values, which allow the stability of the biosphere, have been set in nine areas. Two are very problematic and have far exceeded the tolerable threshold: biogeochemical flows of nitrogen and phosphorus and the integrity of the biosphere, particularly genetic diversity. Two may reveal increasing risks: atmospheric CO_2 concentration and land-use change. Three appear below the thresholds and are tolerable: freshwater use, ocean acidification and stratospheric ozone depletion. Finally, two require further scientific investigation: atmospheric aerosols and new chemical entities.

Environmental limits are global as well as local. This question of scale, the *scalar/interscalar problem*, is essential (Cash *et al.* 2006); scalar referring to the different geographical, temporal, institutional scales, etc., and interscalar for their combination.

First of all, the geographical scale, from the local to the global level. Sustainable development aims to take into account the limits of the planet by changing the mode of development, but is embodied at different levels; national where institutional frameworks are deployed, local where solutions are implemented or even micro-local, those of individual behavior. The density of a resource is decisive, i.e. its economic value in relation to its weight. This ratio determines the distance over which it can be transported economically, whether the resource is of natural origin or from recycling (see Figure 2.2a). Metals will become part of a global market; very heavy construction materials are mobilized at the local level. Two quite different approaches must be considered: a value chain approach for the former and a territorial approach for the latter.

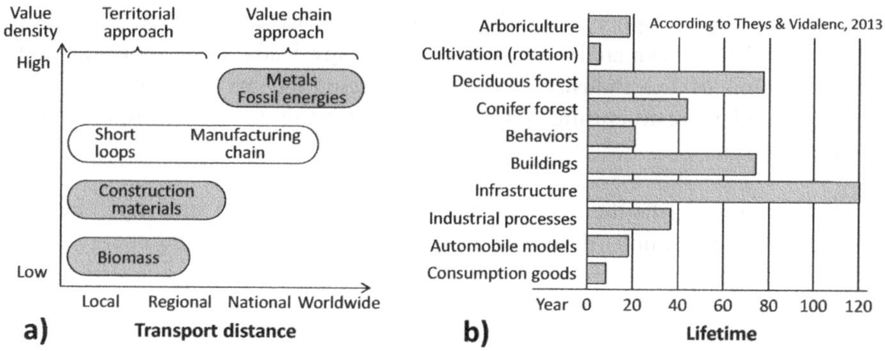

Figure 2.2. *Temporal and territorial dimensions*

The second scale, the temporal scale, in particular the long term, which is undoubtedly the new contribution of sustainable development that considers the control of the effects of current choices on the future, and therefore the entire life span of infrastructure products and their impact on resources and ecosystems. Extending the duration of resource use can lead to a slowdown in overall metabolism and therefore in the growth of resource use. On the other hand, a circular economy policy, focused solely on downstream waste management and the promotion of recycling alone, can lead to an acceleration of metabolism.

These lifespans are very different according to the products, see Figure 2.2b (Theys and Vidalenc 2013), and lead to different strategies for managing them in the context of the circular economy. This can be achieved through the fight against planned obsolescence, by promoting reparability, resilience and frugality, and the simplicity of solutions. It is therefore a question of design, of eco-design.

The structure and geographical size of the market of the circular economy are closely linked to the geographical and temporal material conditions of the resources and their use.

2.3. Methods to address the circular economy

2.3.1. *Historical background: from industrial ecology to circular economy*

It was in an article in *Scientific American* in September 1989 that Robert Frosch and Nicholas Gallopoulos initiated the concept of industrial ecology, proposing an analogy between the concept of the industrial ecosystem and the natural ecosystem. They considered that the industrial system is based on the resources and services provided by the biosphere, which cannot be separated from it (Frosch and Gallopoulos 1989; Erkman 1997). Industrial ecology leads to an understanding of the functioning of the industrial system, its regulation and interaction with the biosphere, and to restructuring it to make it compatible with the functioning of natural ecosystems.

This analogy led to the use of concepts from ecology to industry. "Industrial metabolism", for example, describes all the flows of materials and energy through the industrial system.

At the end of the 1990s, this concept survived through its location in eco-industrial parks and islands of sustainability, where industrial symbioses could develop with waste from one activity becoming the raw material for another. The second approach was the strategy of dematerialization–decarbonization of the economy, including the service economy, or the FE (Erkman 1997).

Historically, industrial ecology has developed a double logic: to be compatible with the biosphere and to adopt the ways in which the biosphere functions, in order to improve this compatibility. In a way, it is a mimicry of the economic system with the functioning of ecosystems, in particular, through the two components: the use of renewable solar energy and the recycling of materials.

The deployment of the circular economy raises two questions, its evaluation and design.

2.3.2. *The lifecycle approach*

The lifecycle approach consists of considering the environmental impacts of each of the stages: raw materials, manufacturing, transportation, use and end-of-life. It aims to minimize the impacts on all these stages, and no longer just one of them: the production or treatment of waste. It makes it possible to avoid pollution transfers from one stage to another.

It is based on an evaluation method, lifecycle assessment (LCA), which makes it possible to carry out a multi-criteria environmental assessment on all the environmental and multi-stage issues of a system (product, service, company or process) over its entire lifecycle. In the case of global value chains that separate production stages geographically, LCA measures the impact on the entire chain by including impacts outside national jurisdictions.

To illustrate this approach to resources, the concept of an ecological backpack was proposed in the 1990s (Hinterberger *et al.* 1997). By measuring material inputs per unit of service (MIPS), it supported the authors' proposed objective of reducing resource use by a factor of 10 (Hinterberger and Schmidt-Bleek 1999). Another proposal, which had more political success, had the same objective: the ecological footprint (Wackernagel and Rees 1996; Boutaud and Gondran 2009). This composite indicator reflects the consumption of resources and waste emissions on the biologically productive surface, to produce these resources and absorb the waste.

This notion of footprint is now used for carbon or water. The carbon footprint considers national emissions and the import/export balance of emissions during the production of goods. This reference is indeed necessary. Emission reduction country commitments only cover emissions that occur within their national territory.

However, in 2015, French national emissions were only 6.6t CO_2e per capita, while the French carbon footprint reached 11 tCO_2e per capita (HCC 2019, p. 34). On a national basis, GHG emissions in France decreased by 19% between 1990 and 2018, still exceeding its commitments by 3.5% (HCC 2019, p. 28). But the French carbon footprint increased by 20% between 1995 and 2015 to the point where imports exceeded emissions on the territory (excluding exports).

This global/local scalar problem is now embodied in new approaches that are still emerging in terms of research. The absolute sustainability approach allows all of these footprints and planetary boundaries to be taken into account. This approach makes it possible to assign the planet's carrying capacity to global value chains and thus to different types of consumption. This approach has been applied, for example, to the impact on biodiversity of a distributor's food portfolio and thus, to individuals and their customers (Wolff *et al.* 2018).

The territorial approach of the circular economy, and of local metabolism, must also integrate the footprint of the territory's consumption at the global level. This is the objective of LCA applied to territories (Albertí *et al.* 2017; Albertí *et al.* 2019a; Albertí *et al.* 2019b).

2.3.3. *Eco-design*

The efficient use of resources, and the reduction of the environmental impacts of products and services, are the main challenges we all face in making the circular economy a reality. They can only be identified from a lifecycle perspective. Meeting these challenges and achieving these objectives requires a systemic approach and strong cooperation between the various actors operating in the supply chain, from producer to final consumer: this is eco-design.

Eco-design is therefore a methodological framework defined in the French standard X30 264, the objective of which is to reduce the environmental impact of products and services at source, from a lifecycle perspective. While it has been suggested that the integration of eco-design into product development processes can reduce environmental impacts by up to 80% (Graedel and Allenby 1995), it is also a source of performance and profitability for companies that implement it. Indeed, the study on the benefits of eco-design, carried out by the Eco-design Center in 2014, found that 96% of companies that implement this approach increase their overall performance (Haned *et al.* 2014). Those that use eco-design as an innovation factor even increase their gross margin in 45% of the cases studied. This leads us to the notion of eco-innovation, defined as an eco-design whose output is a new or improved product or service that offers the user higher usage values, and gives a competitive advantage and economic value to the issuer in the market.

In a controlled and holistic framework of environmental impacts, reducing at source, increasing life expectancy and closing loops are the main eco-design strategies. But eco-design goes beyond this by proposing new business models for companies through lifecycle thinking, such as product-service systems, the economy of sharing or the FE. In this respect, eco-design is at the crossroads of an environmental dimension, assessed throughout the lifecycle and lifespan, and an economic dimension by considering the business model, situating it in the value chain, and thus guiding innovation.

In this logic, eco-design cannot be limited to a component of the circular economy, but a transversal approach to selecting relevant circular economy models or other sustainable consumption and production models, and ensuring their environmental, social and economic coherence.

2.4. Political and normative processes

2.4.1. *The European and French approaches*

In 2015, the European Commission adopted an action plan to accelerate Europe's transition to a circular economy. On this basis, new rules were adopted in

May 2018, setting recycling targets: 60% by 2030 for municipal waste, 55% for plastics, 70% for packaging, 80% for ferrous metals, etc. The European approach is based on the principle of EPR, whereby economic actors (manufacturers, distributors, importers) who place products on the market that generate waste, take charge of all or part of the management of this waste, including separate collection.

The French legislative framework is part of the European framework. The draft anti-waste law for a circular economy, adopted at first reading by the Senate in September 2019, aims to transpose this European Package on the circular economy. It thus focuses the circular economy on the downstream end of the cycle and treatment of waste, and indirectly on the secondary product market, more than on the upstream end and control of resource use. The law will lead to a strengthening of EPR regimes, introducing new obligations for sorting, collection, repair or recycling, and the creation of new waste treatment facilities.

The French doctrine promoted by ADEME is broader, covering three areas based on seven pillars (ADEME 2014):

– offer side economic actors: (1) extraction/exploitation and sustainable purchasing, (2) eco-design (products and processes), (3) industrial and territorial ecology, (4) functional economy;

– consumer demand and behavior: (5) responsible consumption (purchasing, collaborative consumption, use), (6) longer use (reuse, repair);

– waste management: (7) recycling (material and organic).

This pillar design leads to separate approaches. At the legislative level, under the influence of the European principle of EPR, priority is given to the last area, waste. The financing of waste treatment is favored to the detriment of other preventive approaches.

However, eco-design plays a particular role in allowing a choice between these different components, in order to make a real contribution to reducing the use of resources and to coherent ecological management. It plays the role of the "*intelligence*" of the circular economy:

– *Sustainable procurement*: eco-design helps to look for suppliers offering materials or sub-assemblies that are less toxic, renewable, recycled, recyclable, with a lower energy content, "natural", agro-sourced or resulting from reuse.

– *Industrial ecology*: eco-design envisages the control of material flows, inputs or outputs, in particular by considering the resources of a territory and the symbiosis between companies in the same territory.

– *Functional economy* (FE): eco-design can involve economic innovations, such as promoting service and use rather than product ownership. Eco-design can also ensure that the FE takes the environment into account.

– *Responsible consumption*: in addition to the optimal design, with regard to the environment, eco-design gives the consumer credibility and information elements.

– *Extension of the duration of use*: life extension is one of the strategies to reduce the impact of a product, but eco-design makes it possible to consider cases where, for some products, technological developments make new products much more virtuous.

– *Recycling*: finally, end-of-life design is one of the strategies of eco-design, but eco-design considers the limits of this model by comparing it with the environmental impact.

2.4.2. *ISO standardization*

The general objective of the circular economy technical committee, established in 2019 by ISO TC 323, is to develop frameworks, guidance, supporting tools and requirements for the implementation of *circular economy projects*. The proposed deliverables will apply to any organization or group of organizations interested in implementing circular economy projects, including commercial organizations, public services and non-profit organizations. For practical reasons, it will not cover some specific areas of action in the circular economy already covered by existing technical committees, such as eco-design, lifecycle analysis in ISO/TC 207 Environmental management and responsible purchasing (ISO 20400:2017).

The standardization process formalizes exchanges with other standardization processes, thus providing a good overview of the content of the circular economy and its implementation context.

The purpose of standardization is limited to circular economy projects, a term that is rather ambiguous and does not allow the whole issue to be covered. In addition, standards related to the circular economy are based on the concept of continuous improvement, as proposed in ISO 14001. However, management standards for organizations or projects are based on continuous, incremental improvement, pushing to do “the same thing” with less impact on the environment. If the orientation is virtuous, their impact on the resource issue is limited for two main reasons.

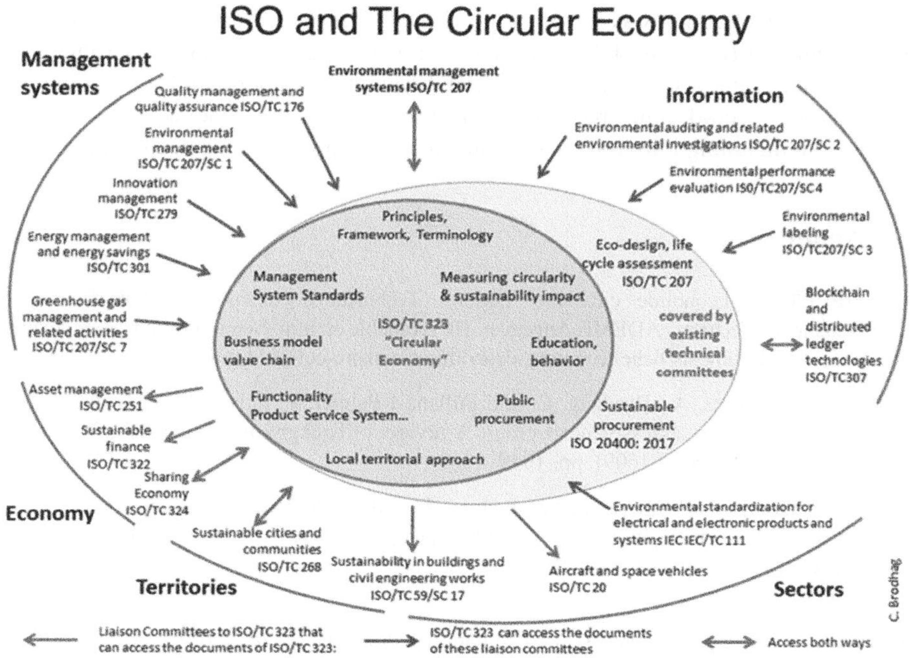

Figure 2.3. *Organization of the standardization process. For a color version of this figure, see iste.co.uk/delchet/circular.zip*

They do not systematically integrate the lifecycle perspective, a principle necessary to avoid any transfer of pollution and the reduction of consumption and impacts. They give no ambition to the projects, while disruptive changes are needed to address resource and environmental impact issues, such as the reduction factor of 10.

2.5. Conclusion

The circular economy is one of the solutions to the challenge of sustainable development, i.e. making development compatible with the limits of the planet.

The various content proposals given to the circular economy are limited by the approaches and rationalities of the institutions that tackle them: the legislative framework favors the downstream end of the lifecycle, waste management and recycling, and standardization favors management systems.

The circular economy can rise to the challenge if its principles and approaches are systematically assessed from an environmental point of view: environmental impact studies and a lifecycle approach, and from an economic and social point of view through eco-design. Its performance objectives must be at the heart of innovation, including initiating disruptions in consumption and production patterns.

2.6. References

ADEME (2014). Economie circulaire: notions. Technical document, Direction Economie Circulaire et Déchets – ADEME Angers, p. 10. Available at: http://www.ademe.fr/sites/default/files/assets/documents/fiche-technique-economie-circulaire-oct-2014.pdf.

Albertí, J., Balaguera, A., Brodhag, C. and Fullana-I-Palmer, P. (eds). (2017). Towards life cycle sustainability assessment of cities. A review of background knowledge. *Science of the Total Environment*, (609), pp. 1049–1063.

Albertí, J., Brodhag, C. and Fullana-I-Palmer, P. (eds). (2019a). First steps in life cycle assessments of cities with a sustainability perspective: A proposal for goal, function, functional unit, and reference flow. *Science of the Total Environment*, (219), pp. 1516–1527.

Albertí, J., Roca, M., Brodhag, C. and Fullana-I-Palmer, P. (eds). (2019b). Allocation and system boundary in life cycle assessments of cities. *Habitat International*, (83), pp. 41–54.

Boutaud, A. and Gondran, N. (2009). *L'empreinte écologique*. La Découverte, Paris.

Cash, D.W., Adger, W.N., Berkes, F., Garden, P., Lebel, L., Olsson, P., Pritchard, L. and Young, O. (eds). (2006). Scale and cross-scale dynamics: governance and information in a multilevel world. *Ecology and Society*, 11(2), p. 8.

Erkman, S. (1997). Industrial ecology: an historical view. *Journal of Cleaner Production*, 5(1–2), pp. 1–10.

Frosch, R.A. and Gallopoulos, N.E. (1989). Strategies for Manufacturing. *Scientific American*, (261), pp. 144–152.

Geldron, A. (2017). L'épuisement des métaux et minéraux : faut-il s'inquiéter ? ADEME.

Graedel, T.E., Allenby, B.R., and Comrie, P.R. (1995). Matrix approaches to abridged life cycle assessment. *Environmental Science & Technology*, 29(3), pp. 134A–139A.

Graedel, T., Buchert, M., Reck, B. and Sonnemann, G. (eds). (2011). Metalstocks and recycling rates, s.l: UNEP.

Haned, N., Lanoie, P., Plouffe, S. and Vernier, M.-F., (eds). 2014. La profitabilité de l'écoconception: une analyse économique, Institut de développement de produits, Pôle écoconception Saint-Etienne, Montreal.

HCC (2019). Agir en cohérence avec les ambitions. Haut Conseil pour le Climat Premier rapport annuel Neutralité Carbone, Paris.

Hinterberger, F., Luks, F. and Schmidt-Bleek, F. (eds). (1997). Material flows vs. 'natural capital': What makes an economy sustainable? *Ecological Economics*, October, 23(1), pp. 1–14.

Hinterberger, F. and Schmidt-Bleek, F. (1999). Dematerialization, MIPS and Factor 10 Physical sustainability indicators as a social device. *Ecological Economics*, 29(1) pp. 53–56.

MEA (2005). *Ecosystems and Human Well-being: Synthesis.* Island Press, Washington, DC.

Oberle, B., Bringezu, S., Hatfield-Dodds, S., Hellweg, S., Schandl, H. and Clement, J. (2019). Perspectives des ressources mondiales. Des ressources naturelles pour l'avenir que nous voulons. ONU Environnement, Nairobi, Kenya.

Rockström, J., Steffen, W., Noone, K., Persson, Å., Chapin, F.S., Lambin, E.F., Lenton, T.M., Scheffer, M., Folke, C., Schellnhuber, H.J., Nykvist, B., de Wit, C.A., Hughes, T., van der Leeuw, S., Rodhe, H., Sörlin, S., Snyder, P.K., Costanza, R., Svedin, U., Falkenmark, M., Karlberg, L., Corell, R.W., Fabry, V.J., Hansen, J., Walker, B., Liverman, D., Richardson, K., Crutzen, P. and Foley, J.A. (eds). (2009). A safe operating space for humanity. *Nature*, 461(7263), pp. 472–475.

Steffen, W., Richardson, K., Rockström, J., Cornell, S.E., Fetzer, I., Bennett, E.M., Biggs, R., Carpenter, S.R., de Vries, W., de Wit, C.A., Folke, C., Gerten, D., Heinke, J., Mace, G., Persson, L.M., Ramanathan, V., Reyers, B., and Sörlin, S. (eds). (2015). Planetary boundaries: Guiding human development on a changing planet. *Science*, 13 February, 347(6223).

Theys, J., and Vidalenc, E. (2013). Repenser les villes dans la société post-carbone. CGDD ADEME, Paris.

Wackernagel, M. and Rees, W.E. (1996). *Our Ecological Footprint: Reducing Human Impact on the Earth.* New Society Publishers, Philadelphia.

WCED (1987). *Our Common Future, (The Brundtland Report).* Oxford University Press, New York.

Wolff, A., Gondran, N. and Brodhag, C. (2017). Detecting unsustainable pressures exerted on biodiversity by a company. Application to the food portfolio of a retailer. *Journal of Cleaner Production*, issue 166, pp. 784–797.

3

Circular Economy: Transformation of the Labor Market and Change in Human Resources Management Practices

The circular economy (CE) is an economic model that combines economic performance with resource conservation. This economic model has no component relative to human resources and human resources management (HRM). The objective of this chapter is to define the impact of the CE on the labor market and how it will transform employment and HR practices. Such reflection helps to understand how the CE can contribute to the renewal of HRM policies and practices. Our central argument is that companies cannot achieve their economic and environmental performance objectives without understanding labor market changes and reviewing their HRM practices. After presenting the impact of the CE on the labor market and jobs, we discuss green human resources management (green HRM) in terms of recruitment, training, and remuneration.

3.1. Introduction

The company has always been at the heart of a conflict between its proximate commercial interests and seemingly more remote societal interests in which its activity is directly involved (Touraine 1969). Extensive growth – which uses a significant amount of resources and leads to many negative externalities (Pigou 1920) such as the depletion of natural resources and the rise of social inequalities (Piketty 2013) – questions the hegemony of capitalism and could lead to its end (David 2016). The traditional productive vision of the company (Fordism) based on capital accumulation is therefore increasingly distant from the reality of the new

Chapter written by David MORIEZ.

societal paradigm: sustainable development (Beaupré *et al.* 2008; Hébert 1990; Waaub 1991).

The definition of "circular economy" is not fixed. It varies by author and between the public and private sectors. The three dimensions of economic and social sustainability are rarely totally addressed. The most comprehensive definition comes from the British Ellen MacArthur Foundation:

> A circular economy seeks to rebuild capital, whether this is financial, manufactured, human, social, or natural (Ellen MacArthur Foundation 2019).

The concept was evoked in France during the 2007 *Grenelle de l'environnement*, and made official with the opening of the *Institut National de l'Economie Circulaire*, at the initiative of MP François-Michel Lambert, on February 6, 2013. The ballot box has recently reflected its importance. In essence, the French Ecological Party (*Europe Ecologie Les Verts*, EELV) rose to third place in the last European elections in May 2019. Therefore, the law on the circular economy, initially presented to MPs on July 10, 2019, arrived before the French National Council for Ecological Transition (*Conseil National de la Transition Ecologique*, CNTE) on June 3 in a renewed social and political context. It lies in the "polluter pays" principle, and it clearly emphasizes the producers' responsibility. The ideal of progress changes fundamentally to move from the primacy of the linear economy (extract – manufacture – consume – throw away) to the primacy of the circular economy (recycle – transform – reuse) (Waaub 1991).

As a result of this paradigm shift, companies are formalizing their environmental practices (Bonnie and Huang 2001). The body of knowledge on the environmentally responsible manufacturing (ERM) process is diversifying and extending to many disciplines (Curkovic and Sroufe 2016) such as green marketing (Peattie 1992), green accounting (Bebbington 2001; Owen 1992), green supply chains (Jabbour and Sousa Jabbour 2016; Lai *et al.* 2010) and green management in general (McDonagh and Prothero 1998).

Though one of the "main obstacles" to the successful deployment of a circular economy in companies is the lack of appropriate human resource management (HRM) (Garcés-Ayerbe *et al.* 2019, p. 1), HRM does not appear in any of the existing models. In January 2020, that of the French Environment and Energy Management Agency (*Agence de l'Environnement et de la Maîtrise de l'Énergie*, ADEME) did not have any social component. Although many international researchers have stressed the importance and challenges of green HRM (Del Brío *et al.* 2007; Govindarajulu Nalini 2004; Jabbour *et al.* 2013; Jabbour and Santos 2008; Madsen and Ulhøi 2001; Massoud *et al.* 2008; Muster and Schrader 2011;

Ramus, 2002, 2001; Renwick *et al.* 2013; Stringer 2010; Wehrmeyer 1996) and the fundamental role of employees in the adoption and deployment of more efficient (i.e. added financial value) and effective (i.e. added quality) environmental practices (Florida and Davison 2001; Garcés-Ayerbe *et al.* 2019; Jackson *et al.* 2011; Renwick *et al.* 2013), the scientific community in France has remained silent. A search of the EBSCO database for peer-reviewed articles in French management sciences published between 2000 and 2019 for which we entered the keywords "management", "green", and "human resources" in the fields "title" and "abstract" did not yield any results.

The objective of this chapter is thus to define the impact of the circular economy on the labor market. More precisely, the employment and the jobs, and how this will change HR practices. Such reflection helps us understand how the circular economy can contribute to the renewal of HRM policies and practices.

We now examine the data on the structure of the labor market in the CE. We see why it is likely to address the dual problem of regional and structural unemployment. From there, we specify the inherent implications for HR practices.

3.2. The circular economy, labor market and jobs

In 2015, the European Commission published its action plan for the circular economy, "Closing the loop", which paves the way for the European Union's transition from a linear to a circular economy. The rationale of which we find in nature, where the lifecycle of the elements never ends but regenerates (European Commission 2015). Regenaration is the EU's strategy for a sustainable, low-carbon, resource-efficient, and competitive economy. An economy that promotes the reuse and recycling of materials and extends the life of the product requires more labor than one based on disposal, i.e. linear. The main reason is that repairing, maintaining, and reconditioning existing final goods takes more time and labor than mining and processing activities (Wijkman and Kristian 2016). In an economy with fewer material resources, the transition from a linear to a circular economy will, therefore, generate more skilled labor (Stahel 2016).

Four recent studies in Europe confirm that the transition to a circular economy will have a positive impact on job creation by 2030:

– a study by the Club of Rome in 2016 indicates that circular economy could create more than 500,000 jobs in France (Wijkman and Kristian 2016);

– a WRAP report in 2015 states that the transition to a circular economy at European level could create between 1.2 and 3 million jobs and reduce unemployment from around 250,000 to 520,000 people (Mitchell and James 2015);

– a study by the European Remanufacturing Network (ERN) in 2015 shows that the remanufacturing activity alone on a European scale could create between 34,000 and 65,000 jobs (Parker *et al.* 2015);

– a study by the Green Alliance and WRAP in 2015 states that the objectives of combating regional unemployment and improving the efficiency of resource use at the national level are linked (Morgan and Mitchell 2015).

According to these four studies, the circular economy is a response to the dual problem of regional and structural unemployment.

3.2.1. *The circular economy: a response to the regionalization of unemployment*

According to INSEE, the French regions with the highest unemployment rates are the Hauts de France (Nord-Pas-de-Calais, Picardie), Occitanie (Languedoc-Roussillon, Midi Pyrénées) and Provence-Alpes-Côte d'Azur (Alpes de Haute Provence, Alpes Maritimes, Bouches du Rhône, Vaucluse). There are three reasons for such regional differences:

– deindustrialization of economic activity;

– automation and robotization of tasks;

– lack of applicants' mobility.

Over the past 50 years, due to technological progress, the share of employment in the industrial sector has fallen by 50%. In other words, the shift from the manufacturing to the service sector destroyed one in two jobs. At the same time, the share of employment in the service sector increased by only 20%. In other words, the service sector has compensated for only two in ten lost jobs from the manufacturing industry (World Economic Forum 2016). Due to the automation of tasks, the industry produces as much with fewer employees. For the industry to maintain its jobs, its growth would have to be more significant. However, according to a report by the Ministry of Economy and Finance in 2017, France is the eurozone country whose share of the manufacturing industry's value-added in gross domestic product (GDP) is the lowest. The balance of foreign trade in manufactured goods reached its lowest historical level in 2017 (-€51 billion) (Aubourg 2017). Regions that have more industrial jobs, such as the Hauts de France, therefore have a higher unemployment rate than the French average due to the deindustrialization of their employment pool since the 1980s (closures of blast furnaces, mines and steel plants). Moreover, job opportunities are not always close to the place of residence of job seekers. According to a recent study by Pôle Emploi, the public employment service in France, the low level of qualifications, the financial cost of mobility (fuel prices) and some extra-professional reasons (lack of driving license and means of transport, young children, rental of social housing and property) explain the attachment to the territory of job seekers, who do not plan, in one case out of five, to extend their geographical search area to more than 50 km from their home (Chamkhi *et al.* 2019).

Research is struggling to accurately measure the share of unemployment attributable to this geographical mismatch between labor supply and demand. France Stratégie, administratively called the *Commissariat Général à la Stratégie et à la Prospective*, an institution close to the Prime Minister, estimates that geographical mismatches in France represent between 1 and 2.5 percentage points of the unemployment rate (Gilles and Sode 2019). No research has been conducted in France to measure the impact of circular economy on the territorial redistribution of employment. Still, the Green Alliance and WRAP study in 2015 for Great Britain shows relevant results for all developed countries. The level of employment concentration depends on the nature of the circular economy activity (Morgan and Mitchell 2015).

Activity	Level of employment concentration	Area of concentration
Reuse	Very low	Scattered all over the country.
Closed-loop recycling	High	Close to manufacturing sites, supply chains, and logistics.
Open-loop recycling	Moderate	Close to raw materials, markets, and major ports.
Bio-refining	High	Near major ports, consumer industries, manufacturing sites, major urban areas, and raw material sources.
Remanufacturing	Very high	Near manufacturing sites, transport networks, major urban areas, and abroad.
Servicing	Low	The head office can be located in Paris. The back office and service jobs can be anywhere in France or abroad.

Table 3.1. *Circular economy activities and employment concentration (adapted from Morgan and Mitchell 2015)*

The circular economy will affect employment mapping as we know it today (see Table 3.1). In other words, the circular economy can reduce regional unemployment mismatches (Mitchell and James 2015, p. 4).

Reconditioning, bio-refining, and closed-loop recycling (use of recycling material for the same purpose, e.g. recycling of a PET bottle into a PET bottle, recycling of packaging glass into packaging glass, and recycling of road mixes in the manufacture of new mixes) are high concentration activities. With the new "polluter-pays" principle, high concentration activities must optimize their flows. They cannot relocate to remote places. They need to be physically present (Abitbol *et al.* 2014).

The reuse, service (maintenance and repair), and open-loop recycling (use of recycled material for a different destination, e.g. recycling a PET bottle made of polar fiber and recycling paper into an insulation product) are low concentration activities. The circular economy requires greater geographic proximity between producers and users. In other words, the new circular economy law is much more comprehensive and more ambitious than the first version circularized in January 2019 by the Ministry of Ecological and Solidarity Transition (*Ministère de la transition écologique et solidaire*, MTES). It requires different sectors to identify and map their maintenance and repair services so that most consumers can find them easily and nearby. Reuse, servitization, and recycling activities in an open-loop will, therefore, lead to the spread of economic activity beyond the usual areas of consumption or capillarization.

To sum up, transitioning from a linear to a circular economy will cause both the polarization and capillarization of employment. This combined phenomenon is likely to solve the problem of unemployment regionalization. Very high and high employment concentration activities (i.e. remanufacturing, closed-loop recycling, bio-refining) will maintain employment in agglomerations and suburbs where it is usually concentrated. At the same time, very low, low, and moderate (i.e. reuse, servitization, open-loop recycling) employment concentration activities will disperse employment throughout the territory. Open-loop recycling, servitization and reuse activities will therefore assist in revitalizing regions where the unemployment rate is higher (Hauts de France, Occitanie, and Provence-Alpes-Côte d'Azur).

Transitioning from a linear to a circular economy will also lead to the cross-sectoral redistribution of jobs and the convergence of skills. It could, therefore, address the problem of structural unemployment.

3.2.2. *The circular economy: a response to structural unemployment*

Structural unemployment is the result of the qualitative mismatch between supply and demand in the labor market. The evolution of qualifications due to technical developments leads to a decline in the employability of a part of the working population that can no longer find jobs corresponding to its qualifications.

The OECD and the European Commission, which are the two leading institutions that regularly publish estimates of the structural unemployment rate, estimate that the structural unemployment rate in France is 9%. While this estimate may seem high, the unemployment rate in France has not fallen permanently below 7.5% for 30 years, regardless of the level of growth or the orientation of French economic policy (Gilles and Sode 2019). The Fourth Industrial Revolution (i.e. dematerialization, object interconnection, and artificial intelligence) will have destroyed 7 million white-collar jobs by 2020 (Davos Forum 2016). These losses will only be partially offset by the creation of 2 million new jobs, mainly in specialized fields such as the computer sciences, mathematics, and engineering (World Economic Forum 2016). At the same time, according to a 2016 Club of Rome study, the circular economy will benefit some branches of the primary sector, such as agriculture and forestry, with the increased demand for biofuels. Demand for renewable power plants, and the installation and construction services inherent in the deployment of these plants will also increase. The manufacture of durable consumer goods, in the economic sense of the term (i.e. those with a life span of more than three years, such as furniture, automobiles, household appliances and consumer electronics, sports equipment and toys), will decrease. However, the number of companies remanufacturing, maintaining, repairing, and recycling these same goods will increase (Wijkman and Kristian 2016). The combined effect of the Third and Fourth Industrial Revolutions and transition from a linear economy to a circular economy would, therefore, invite us to revisit Schumpeter's (1942) theory of creative destruction:

> The same process of industrial mutation (…) that incessantly revolutionizes the economic structure from within, whilst incessantly destroying the old one, and incessantly creating a new one (Schumpeter 1943, p. 83)

With the Third Industrial Revolution at the end the of the 20th Century (Rifkin *et al.* 2012), the new information and communication technologies, the Fourth Industrial Revolution (Schwab 2017), and the advent of artificial intelligence, 3D printers, and cobotics (i.e. human–robot collaboration), there is more into the transition from a linear to a circular economy than destroying jobs in fossil fuels and creating new ones in renewable energies. The jobs created within the framework of the circular economy covers a wide range of skills. The transition from a linear to a circular economy requires production agents (PA), employees, technicians, and managers (ETAM) as well as engineers and managers (EM) at different skill levels: "low" for recycling and reuse activities, "medium" for remanufacturing activities and "high" for bio-refinery activities (Morgan and Mitchell 2015). Highly-skilled workers are necessary to create and develop new production loops. Lower-skilled workers are necessary to operate and maintain production loops (Deboutière and Georgeault 2015). According to France Stratégie, the development of activities to repair used products, reuse, or recycle waste generates about 25 times more jobs than landfilling this waste (Jolly and Douillard 2016). The transition from a linear to a circular economy will, therefore, lead to an increase in the employability of the primary (agriculture, forestry) and secondary (industries repairing, reusing, recycling) sectors. The employees in the primary and manufacturing sectors will undergo a significant change in their technical skills. However, relatively few new jobs will emerge. More precisely, the transition from a linear to a circular economy will lead to a convergence of transversal practices (i.e. eco-responsible) for all "green" and "greening" occupations.

3.2.3. *The circular economy: "green jobs" and "greening jobs"*

The terminology used to describe jobs related to the environment is varied, unstable, and does not always cover the same fields of activity. It is, therefore, impossible to exhaustively list the professions concerned by the transition from a linear to a circular economy. We invite the reader to go and look at our documentary selection on green economy jobs at the end of this chapter (see section 3.5. Appendix 1).

The literature generally distinguishes between "green jobs" and "greening jobs". "Green jobs" have an environmental purpose. They measure, prevent, control, and correct the negative impacts and damage of human activity on the environment. "Greening jobs" do not have an environmental purpose. Their technical base remains unchanged, but they are incorporating new skill elements to take into account the environmental dimension of their actions.

In terms of green jobs, we can mention the new profession of master composter for which the French Environment and Energy Management Agency (ADEME) has created the reference system. By 2025, in France, all vegetable waste will have to be collected separately in buildings, neighbourhoods, and businesses. The master composter is the point of contact of one or more composting sites. He/she raises public awareness of the challenges of plant waste and provides training in the collection and treatment of this waste. There is also Michael Page's (2010) study of the functions and wages of green jobs (see section 3.6. Appendix 2). Although this study is old, it remains enlightening because it comes from an empirical and cross-referenced analysis of three sources of information: application files, actual data in terms of functions and salaries, and the history of more than 20 years' presence on the recruitment market. It makes it possible to highlight 17 reference professions divided into seven sectors. Finally, other professions, such as those resulting from the convergence of the energy, waste, and security sectors, will emerge (see section 3.7 Appendix 3).

With greening jobs, employees will have to develop long-term sustainable and ecological thinking to carry out their activities. All disciplines are concerned, as evidenced by the research inherent in green marketing (Peattie 1992), green accounting (Bebbington 2001; Owen 1992), green supply chains (Lai *et al.* 2010), and green management in general (McDonagh and Prothero 1998). For example, at the marketing level, new types of trade are emerging, such as bulk trade. The sale of local products and the implementation of short circuits are developing. Sellers will have to be able to integrate the notions of circular economy into their sales arguments and marketing managers into their marketing plan (product, price, communication, distribution). At the banking and insurance level, the circular economy will lead to changes in the supply of insurance and financial products, and the terms and conditions of bank loans and hedging agreements. In terms of research and development, engineers, designers, and creators will have to specialize in eco-design. Their job will be to design a product by making it as environmentally friendly as possible at each stage of its lifecycle. Both green and greening jobs will, therefore, benefit from increased transversality and employability. They will become much more attractive.

From 1997 to 2007, according to Studyrama (2010), a French site specializing in student training and orientation, enrolment in sustainable development training grew by 3% per year, while at the same time, total enrolment in training increased by only 1.1% per year. Until 2010, young graduates from these courses were therefore leaving at a higher rate than vacancies or job creation and it took them on average longer to get a job (Roger 2010). In May 2019, a study by the French General Commission for Sustainable Development for the Ministry of Ecological and Solidarity Transition showed that the number of jobs in the environment sector increased by 5.4% between 2016 and 2017. It reached 465,450 full-time equivalent jobs or 1.8% of total domestic employment in France. Over the same period, in proportion to the economy as a whole, the 24,000 new "green jobs" represented a +1% increase in job creation in 2017 (*Commissariat Général du Développement Durable* 2019). The spillover effect which indicates that technological progress improves productivity (the ratio between the quantity of production and the number of hours needed to produce it) and generates a linear transfer (spill) of jobs from one economic sector to another (Sauvy and Hirsch 1980) is therefore to be nuanced. The spillover of jobs associated with the transition from a linear to a circular economy is multisectoral and multidisciplinary.

At the end of this section on the impact of the circular economy on the labor market and jobs, we note that the transition from a linear to a circular economic model will lead to a profound transformation of the labor market. There will be a polarization of remanufacturing, bio-refining, and closed-loop recycling activities, and a capillarization of reuse, service, and open-loop recycling activities. If relatively few new occupations emerge, the "green" and "greening" effects of the circular economy will lead to a general increase in the need for workforces specific to all economic sectors, skill levels, and disciplines, which will help to combat regionalization and structural unemployment.

The development of appropriate human resources (HR) practices is an essential prerequisite for a coherent approach. However, there is a resource traditionally overlooked in the literature on the circular economy: the human resource. In the second section, we therefore discuss the impact of the circular economy on HR practices.

3.3. The circular economy and human resources management practices

There is a fundamental movement that began in the 1980s: the reflexivity of work. The reflexivity of work implies that each employee increasingly needs to align his/her life and work values (Moriez and Grima 2019). More precisely, the decline of extrinsic (i.e. organizational) values in favor of intrinsic (i.e. individual) values radically changes the relationship to work (Ester *et al.* 2006; Riffault and Tchernia 2002, pp. 63–80; Tchernia 2005). Under the influence of this increased reflexivity of work, the definition of professional success is changing. The linear criteria of the position (salary, seniority, career progression, etc.) weaken as the circular criteria of the job (sense of work, environmental awareness, etc.) intensify. Happiness at work is, therefore, no longer limited to the material conditions of performance as formalized in the contract of employment, but implies less formalized principles such as being able to contribute to the common good.

3.3.1. *Green human resources management (green HRM)*

The integration of sustainable development into human resources management policies is called green human resources management (green HRM) or "green HR" (Dutta 2012; Margaretha and Saragih 2013). It has an impact on the entire HR function from the employee's integration to his/her departure.

A literature review indicates that there are five main issues for green HR (Cherian and Jacob 2012):

– a reduction in the company's environmental impact;

– an improvement in the company's brand image;

– better attractiveness (number of applications) of the company;

– increased employee retention;

– increased performance (competitiveness and productivity).

Research shows that employee involvement in environmental actions contributes to the "effectiveness" of the company's overall sustainable management policy (Bangwal and Tiwari 2015, p. 49). This effectiveness depends on the degree to which employees adhere to the company's sustainable development values (Lok and Chin 2019). In other words, the quality of results depends on employees' willingness to collaborate (Collier and Esteban 2007). More specifically, off-role or discretionary behaviors are as important as in role or non-discretionary behaviors: "both contribute to organizational results through value creation" (Dumont *et al.* 2017, p. 615). Not smoking in a public building is in-role or non-discretionary behavior. It is not done at the employee's "discretion". It is the result of a collective rule. It is a behavior dictated by law, the employment contract, and the internal regulations. However, off-role or discretionary behavior is at the employee's discretion. It is the result of an individual initiative. It is a free, chosen behavior. For example, turning off the computer and switching the lights off at the end of a working day in a company that does not explicitly encourage this behavior. Therefore, whether a behavior is in-role or not depends on the company (Paillé and Boiral 2013) and its climate (Dumont *et al.* 2017). Due to the interactions between professional and private life, some authors go so far as to argue that the role of human resources should be to encourage responsible behavior at work and outside work according to a logic of ecological balance between personal and professional life (green work–life balance) (Muster and Schrader 2011). There is, therefore, a link between individual engagement (sustainability engagement) and collective green practices in companies (Kim *et al.* 2019; Lok and Chin 2019). The literature on recruitment, training, and remuneration in particular highlights this link (Mohd Yusoff *et al.* 2018).

3.3.2. *Green human resources management practices*

3.3.2.1. *Recruitment and selection*

Research shows that candidates are often aware of the company's sustainable development policy and that their decision to accept a job depends on this policy (Stringer 2010). The successful implementation of a green human resources management policy woud thus begin with appropriate green recruitment.

Green recruitment and selection depends on the low-carbon steps, from the definition of the position to the integration of the employee according to a general orientation of "zero paper, zero fuel" (Saini and Shukla 2016). In other words, green recruitment is the recruitment of talents who are familiar with conservation and sustainability concepts (Bangwal and Tiwari 2015). More specifically, green recruitment depends on new entrants' knowledge of practices that contribute to the effective implementation of a sustainable performance management policy in the company (Wehrmeyer 1996).

Green recruitment will begin with an advertisement that will be published online. As for Patagonia, a Californian company specializing in eco-designed mountain sports and surf clothing, the HR department emphasizes the importance for the candidate to demonstrate environmental responsibility when applying. It encourages online applications. In the pre-selection phase, interviews are conducted by telephone or video to reduce paper and fuel consumption related to travel. During the selection phase, the job descriptions used to formulate questions when interviewing candidates include responsible behavior qualities (Saini and Shukla 2016). HR managers can test these qualities during the pre-selection phase too. They can evaluate the degree of consistency between the candidate's and the company's eco-responsible values. The effectiveness of the implementation of a sustainable performance management policy, therefore, depends on the job descriptions that both define and serve as a benchmark of sustainability in the workplace (Mandip 2012).

3.3.2.2. *Training and development*

The second most crucial lever of green HRM is training: "innovation and training are essential issues in terms of jobs. (They) are highlighted by all the studies on ecological transition" (Deboutière and Georgeault 2015, p. 38). The French Ministries of Ecological and Solidarity Transition and Economy and Finance (2018) recall that the transition from a linear economy to a circular economy "will lead to job transfers (…) [and that] a vigorous training effort facilitating these transfers must be undertaken" (Ministry of Ecological and Solidarity Transition and Ministry of Economy and Finance 2018, p. 4). In other words, the successful deployment of green initiatives in companies requires both technical and managerial skills (Callenbach *et al.* 1993). Technical skills refer to green jobs whose business skills are essential to measure, prevent, control, and correct the negative impacts and damage of human activity on the environment. Managerial skills concern all green and greening jobs since it is a question of taking into account the environmental dimension of actions to change behavior.

Involving employees in solving environmental problems in order to stop the decline in knowledge (expertise), know-how (skills), and attitudes (qualities) related to sustainable development is fundamental (Zoogah 2011). In May 2019, Nestlé joined forces with CleanCup, a young Lyon start-up, to change the way its employees drink. CleanCup's mission is to eliminate the use of disposable cups. The operation of the machine is simple: the user puts one euro deposit, and collects a clean glass that bears the company logo. Once he/she has finished drinking, he/she puts his/her glass back into the machine, recovers his/her deposit, and the machine automatically washes the glass on site (CareNews 2019). Beyond the technical and collective training formalized in the training plans, the success of a green training policy would, therefore, depend on participative and empowering management practices that encourage eco-intrapreneurship in the workplace.

3.3.2.3. *Compensation and benefits*

Employees assume their environmental responsibility better when it is accompanied by a compensation and benefits policy in response (Forman and Jorgensen 2001). Companies can use monetary and non-monetary rewards to encourage the acquisition and demonstration of green skills at work. Monetary rewards include, for example, bonuses based on green objectives or results. Non-monetary rewards include, for example, gifts, internal communication actions, and the link between opportunities for advancement and the employee's level of green initiatives (Ramus 2002). These initiatives go beyond recycling actions. They may involve accepting flexible hours, thus contributing to the reduction of pollution peaks, or volunteering for telework (Jackson *et al.* 2011). Compensation and benefits are, therefore, other powerful tools to support a green HRM policy and facilitate the achievement of the company's sustainable development objectives (Milliman and Clair 1996).

3.4. Conclusion

The objective of this chapter was to define the impact of the CE on the labor market and how it will transform employment and HR practices.

At the end of section 3.2, we saw that the transition from a linear to a circular economy would lead to a profound transformation of the labor market under the combined effects of both the polarization and capillarization of activities induced by the CE. If few new occupations emerge, the "green" and "greening" effects of the CE will contribute to combating the regionalization and structural nature of unemployment.

Section 3.3 shows that the challenges of green HRM are numerous. The main levers for action are recruitment, training, and compensation. Efforts will be in vain without the coherence of technical and managerial actions, and the individual commitment of employees encouraged to become eco-intrapreneurship agents.

3.5. Appendix 1

Documentary selection on green economy jobs:

Title, publisher, year of publication	Description
Guide des formations aux énergies renouvelables (Guide to renewable energy training) 2018–2019, Observ'ER, 2018	This guide includes 220 training courses on renewable energy and eco-construction from BAC+2 to BAC+5 generalists or specialists.
Les métiers de l'environnement et du développement durable (Environment and sustainable development jobs), ONISEP, 2015	This guide presents 31 jobs in risk prevention, environmental protection, and pollution treatment. The courses of study and diplomas are detailed and accompanied by an address book of educational institutions and data on employment trends. It contains highly informative testimonies from professionals.
Les emplois de demain (écolos, collaboratifs, solidaires, innovants) (Tomorrow's jobs (eco-friendly, collaborative, solidarity, innovative)), Terra Eco, 2015	This special edition presents 100 jobs of the future in 10 key sectors (energy, construction, nature protection, cities, transport, pollution, waste, eco-design, alternative sales, information, and communication) and through 50 training courses. It contains highly informative testimonies from professionals.
Le marché de l'emploi de l'économie verte (The green economy labor market), Working Paper No. 44, Observatoire National des Emplois et des Métiers de l'Economie Verte, 2018	Job figures, lists of activities related to green growth, and "green" or "greening" jobs.
Etude sur les mutations des emplois et des métiers des secteurs de la banque, de la finance et de l'assurance dans l'économie verte (Study on the changes in jobs and professions in the banking, finance and insurance sectors in the green economy), Ernst & Young, 2011	Carried out as part of the national mobilization plan for jobs and professions in the green economy, this study diagnoses the links between the green economy and the financial industry. Based on this diagnosis, an overview of the current and future structure of employment in the banking, insurance, and finance sectors is presented, completed by avenues and actions in the field of employment and training policy to better support the expected changes.
Etude sur les mutations des emplois et des métiers des secteurs du commerce et de la distribution dans l'économie	This study has three objectives: identify the specific impacts of the green economy on the trade and distribution sectors; describe the related changes in employment and jobs; and develop courses of action making it possible to

verte et équitable (Study on the changes in employment and jobs in the trade and distribution sectors in the green and fair trade economy). Final report, Ambroise Bouteille et associés, 2012	deduce policies and support tools in human resources management and training.

Table 3.2. *A documentary selection on green economy jobs*

3.6. Appendix 2

APEC green economy executive employment study (2010):

Sector	Career	Mission statement
Renewable energies	Renewable energy project manager	In charge of creating and managing a wind, solar, biomass, or biogas power generation park, through the development of new projects, the purchase of existing parks or through co-investment with other players.
	Commercial renewable energies (Renewable energy sales engineer building, or Sales engineers for renewable energies and public procurement)	In charge of marketing renewable products/solutions (heating, roofing, insulation, energy) to distributors, traders, generalist or specialized artisans, private individuals or local authorities.
	Energy efficiency engineer for buildings	In charge of carrying out energy audits (pre-diagnosis, energy diagnosis, thermal simulation, specific studies of energy savings), coordinating, implementing, and developing an energy performance contract offer for customers based on regular technological and regulatory monitoring.

Environment	Environmental law lawyer	In charge of assisting (monitoring, audit, management, negotiation, obtaining authorizations, prevention, liaison with the DRIRE, prefectures, and environmental protection associations) in the creation, operation, and closure of industrial and agricultural activities and disputes (administrative, judicial, civil and criminal).
	River hydraulics project manager	In charge of flood forecasting and management, flood control and planning, runoff control, flood inventories and prevention plans (PPRI), river development (banks and infrastructure) and wildlife conservation and flora.
	Project manager for polluted sites and soils	In charge of examining the different types of pollution and their consequences in order to secure land transactions and propose techniques for envrionmental clean-up and site rehabilitation.
	HQE (High Environmental Quality) project manager	In charge of defining and communicating to the various stakeholders (customer, design office and staff), the energy and environmental objectives in compliance with regulations. Also in charge of thermal and energy simulations.
	Carbon footprint expert	In charge of collecting data in order to establish a diagnosis, inform staff of the carbon footprint approach and propose an action plan to reduce CO_2 emissions.
Sustainable development	Sustainable development manager	In charge of defining the company's sustainable development policy and communicating it internally and externally; setting up monitoring indicators relating to sustainable development objectives; raising

		awareness, uniting and motivating stakeholders to integrate sustainable development into their strategic decisions; promoting the integration of sustainable development into the commercial offer and supporting CODIR in its reflection on the necessary transformation of the company in light of current societal changes.
	Diversity and CSR (corporate social responsibility) manager	In charge of defining a policy on solidarity, diversity, professional equality and disability applied to HR activities (recruitment, mobility, promotions, recognition, etc.), and to manage training and programs on diversity and discrimination.
	Sustainable development consultant	In charge of advising and assisting clients in the fields of eco-design, lifecycle and technical-economic analyses, environmental information communication, and sustainable development strategy.
Fairtrade and sustainable development purchasing	Buyer/head of fairtrade product	In charge of promoting the benefits of fairtrade and launching new products, he/she accompanies the product from its conception or selection, in a logic of strong partnership with fairtrade producers.
	Sustainable development buyer	Transversely, as a purchasing advisor on goods and services to all the company's departments, he/she is in charge of helping them achieve sustainable development objectives in order to achieve a comparable level of sustainable development. Vertically, he/she is in charge of the administrative and technical structuring of a call for tenders (market access, determination of sustainable development criteria, etc.), and the selection of suppliers based on the assessment of their environmental impact.

Living environment	Sound project manager	In charge, within a company or a research and consulting firm, of carrying out site studies, recording and measuring the sound environment, proposing acoustic solutions adapted to the customer's needs, drawing up estimates, carrying out studies on the sizing of acoustic protection, and carrying out noise maps, seeking subcontractors and partnerships, monitoring the production, and contributing to the development and marketing of acoustic products distributed by the company.
Waste products	Incineration plant operations manager	In charge of leading his/her team by training and developing the skills of his/her employees, managing his/her activity following the company's objectives, regulations and procedures, with the implementation and development of tools and indicators, analyzing and managing his/her operating account following the objectives set, applying the regulations in force and the company's environmental policy, and ensuring the safety policy towards his/her employees. In the case where the heat produced by the incineration is recovered to supply the boiler room of housing or public buildings, he/she is the N+1 of the thermal engineering engineer, specialist in heat management, and energy savings.
	Waste treatment operations manager	In charge of managing landfills designed and implemented by the waste treatment engineering department, he/she manages the treatment processes: landfilling, incineration, and composting.

Socially responsible investment funds	Specialized manager for socially responsible investment	In charge of optimizing the performance of funds, rejecting stocks in the portfolio of companies that do not respect the moral order, guaranteeing the payment of part of the fund's outperformance to charities, ensuring reporting (written and oral) to the manager and to institutional clients in the context of discretionary management and possibly to individual clients in the management of company mutual funds, and contributing to the implementation of statistical tools and financial means aimed at improving the management process.

Table 3.3. *The 17 reference green executive jobs and the 7 sectors (adapted from APEC 2010)*

3.7. Appendix 3

Prospective study of CNIM's green jobs by 2030 (CNIM 2017):

Business line	Description
Expert in rudological sourcing	In 2030, waste will be a resource of choice, which must be identified, recovered, and recycled optimally and securely. The expert in rudological sourcing will be the specialist in this new value chain.
Expert in industrial scalometry	By 2030, factories will be more energy-efficient, (i.e. cleaner and more flexible). Production tools and processes, which are highly digitalized, will adapt to demand in real-time. The scalometer will act as an advisor to adapt production to the needs as well as to the evolution of the urban and regulatory environment.
Circular ecologist	By 2030, urban dwellers will represent nearly two-thirds of the world's population. Waste and residues from human activity will be transformed into resources. As the conductor of the energy recovery orchestra, the circular ecologist will supervise all flows, from their collection at the source and their recovery to their reintegration into the virtuous loop of the circular economy.
Micro-energy architect	By 2030, with multiple and widespread energy sources, monitoring will become essential. Energy recovery and redistribution will be refined and secured through complex sensor networks and control solutions. The micro-energy architect will design these networks and

	will be able to reconcile the fluctuating capacities of renewable energy sources with the needs of consumers who have also become energy producers.
Projection emergency physician	By 2030, environmental or geostrategic emergencies will occur more frequently. To respond to disaster situations, States as well as public and private institutions will deploy comprehensive logistical resources. In the living bases, energy will be self-produced and drinking water treatment plants will treat the water. The projection emergency physician is there to plan interventions in distant areas, aiming to set up fully autonomous bases.

Table 3.4. *Circular economy trades in 2030 (adapted from CNIM 2017)*

3.8. References

Abitbol, L., Blavot, C., Duret, B., Georgeault, L., Mat, N., Rault, M. and Valluis, C. (eds) (2014). Références : écologie industrielle et territoriale : le guide pour agir dans les territoires. Guide, Commissariat Général au Développement Durable, France.

Aubourg, A. (2017). Analyses : l'industrie manufacturière en 2017. Ministère de l'Economie et des Finances, Ivry-sur-Seine.

Bangwal, D. and Tiwari, P. (2015). Green HRM – A way to greening the environment. *IOSR J. Bus. Manag.*, 17, pp. 45–53.

Bebbington, J. (2001). Sustainable development: A review of the international development, business and accounting literature. *Accounting Forum*, 25(2), pp. 128–157 [Online]. Available at: https://doi.org/10.1111/1467-6303.00059.

Bonnie, F.D. and Huang, S.-C. (2001). Achieving sustainability through attention to human resource factors in environmental management. *Int. J. Oper. Prod. Manag.*, 21, pp. 1539–1552.

Callenbach, E., Capra, F., Goldman, L., Lutz, R. and Marburg, S. (eds) (1993). *Eco-Management: The Elmwood Guide to Ecological Auditing and Sustainable Business.* Berrett-Koehler, San Francisco.

CareNews (2019). Zoom sur Cleancup, la machine qui veut changer notre façon de boire [Online]. Available at: https://www.carenews.com/fr/news/13839-cnj-zoom-sur-cleancup-la-machine-qui-veut-changer-notre-facon-de-boire.

Chamkhi, A., Garcia, S., Lim, Y.-M. and Poujouly, C. (eds) (2019). Éclairages et synthèses : la mobilité géographique des demandeurs d'emploi : freins et leviers 51. Report, Pôle Emploi, Paris.

Cherian, J.P. and Jacob, J. (2012). A study of green HR practices and its effective implementation in the organization: A review. *Int. J. Bus. Manag.*, 7, p. 25 [Online]. Available at: https://doi.org/10.5539/ijbm.v7n21p25.

Collier, J. and Esteban, R. (2007). Corporate social responsibility and employee commitment. *Bus. Ethics Eur. Rev.*, 16, pp. 19–33 [Online]. Available at: https://doi.org/10.1111/j.1467-8608.2006.00466.x

Curkovic, S. and Sroufe, R. (2016). A literature review and taxonomy of environmentally responsible manufacturing. *Am. J. Ind. Bus. Manag.*, 6, pp. 323–346 [Online]. Available at: https://doi.org/10.4236/ajibm.2016.63030.

David, A. (2016). Jeremy Rifkin, 2014, La nouvelle société à coût marginal zéro, Editions Les Liens qui Libèrent, Paris. *RIMHE Rev. Interdiscip. Manag. Homme Entrep.*, 20, pp. 105–112 [Online]. Available at: https://doi.org/10.3917/rimhe.020.0105.

Dumont, J., Shen, J. and Deng, X. (2017). Effects of green HRM practices on employee workplace green behavior: The role of psychological green climate and employee green values: Effect of green HRM on employee workplace green behavior. *Hum. Resour. Manage.*, 56, pp. 613–627 [Online]. Available at: https://doi.org/10.1002/hrm.21792.

Dutta, S. (2012). Greening people: A strategic dimension. *ZENITH Int. J. Bus. Econ. Manag. Res.*, 2, pp. 143–148.

Ellen MacArthur Foundation (2019). Economie circulaire. WWW Document [Online]. Available at: https://www.ellenmacarthurfoundation.org/fr/economie-circulaire/concept (accessed August 31, 2019).

Ester, P., Braun, M. and Vinken, H. (2006). Eroding work values? In *Globalization, Value Change, and Generations: A Cross-National and Intergenerational Perspective*, Ester, P., Braun, M., Mohler, P.P. (eds). pp. 89–113, Brill, Leiden.

Gilles, C. and Sode, A. (2019). Chômage conjoncturel et chômage structurel. Report, France Stratégie, département Economie-Finances, Paris.

Jackson, S.E., Renwick, D.W.S., Jabbour, C.J.C. and Muller-Camen, M. (eds) (2011). State-of-the-art and future directions for green human resource management: Introduction to the special issue. *Z. Fuer Pers. Ger. J. Res. Hum. Resour. Manags,* 25, pp. 99–116 [Online]. Available at: https://doi.org/10.2307/23279428.

Jolly, C. and Douillard, P. (2016). L'économie circulaire, combien d'emplois ? Report no. 46, France Stratégie, département Economie-Finances, Paris.

Kim, Y.J., Kim, W.G., Choi, H.-M. and Phetvaroon, K. (eds) (2019). The effect of green human resource management on hotel employees' eco-friendly behavior and environmental performance. *Int. J. Hosp. Manag.*, 76, pp. 83–93 [Online]. Available at: https://doi.org/10.1016/j.ijhm.2018.04.007.

Lai, K.-H., Cheng, T.C.E. and Tang, A.K.Y. (eds) (2010). Green retailing: Factors for success. *Calif. Manage. Rev.*, 52, pp. 6–31 [Online]. Available at: https://doi.org/10.1525/cmr.2010.52.2.6.

Lok, S.Y.P. and Chin, T.L., (2019). Linking Green HRM Practices and Employee Sustainability Engagement. *Glob. Bus. Manag. Res.*, 11, pp. 151–156.

Mandip, G. (2012). Green HRM: People management commitment to environmental sustainability. *Res. J. Recent Sci.*, 1, pp. 244–252.

Margaretha, M. and Saragih, S. (2013). Developing new corporate culture through green human resource practice. In *The 2013 IBEA, International Conference on Business, Economics and Accounting*, Bangkok, Thailand.

McDonagh, P. and Prothero, A. (1998). *Green Management: A Reader*. International Thomson Business Press, Stamford.

Milliman, J. and Clair, J. (1996). Best environmental HRM practices in the US. In *Greening People: Human Resources and Environmental Management*, Wehrmeyer, W. (ed.). pp. 49–73, Greenleaf Publishing Limited, London [Online]. Available at: https://doi.org/10.9774/GLEAF.978-1-909493-00-1_4.

Mitchell, P. and James, K. (2015). Economic growth potential of more circular economies. *WRAP*, September

Mohd Yusoff, Y., Nejati, M., Kee, D. and Amran, A. (eds) (2018). Linking green human resource management practices to environmental performance in the hotel industry. *Glob. Bus. Rev.* [Online]. Available at: https://doi.org/10.1177/0972150918779294.

Morgan, J. and Mitchell, P. (2015). Employment and the circular economy – Job creation in a more resource-efficient Britain. Report, Green Alliance & Wrap, London.

Moriez, D.C. and Grima, F. (2019). Développer un inventaire des méta-valeurs individuelles. Rev. *Int. Psychosociologie Gest. Comport. Organ.*, 25.

Muster, V. and Schrader, U. (2011). Green work-life balance: A new perspective for green HRM. *Ger. J. Hum. Resour. Manag.*, 25, pp. 140–156 [Online]. Available at: https://doi.org/10.1177/239700221102500205.

Owen, D. (ed.) (1992). *Green Reporting: Accountancy and the Challenge of the Nineties*. Chapman & Hall, London.

Parker, D., Riley, K., Robinson, S., Symington, H., Tewson, J., Jansson, K., Ramkumar, S., and Peck, D. (eds) (2015). Remanufacturing market study for Horizon 2020. Report, European Remanufacturing Network.

Peattie, K.J. (1992). *Green Marketing*. The M & E Handbook Series, Pittman, London.

Piketty, T. (2013). *Le capital au XXIe siècle, les livres du nouveau monde*. Le Seuil, Paris.

Ramus, C.A. (2002). Encouraging innovative environmental actions: What companies and managers must do. *J. World Bus.*, 37, pp. 151–164 [Online]. Available at: https://doi.org/10.1016/S1090-9516(02)00074-3.

Riffault, H. and Tchernia, J.-F. (2002). Les Européens et le travail : un rapport plus personnel. Report, Futuribles, Paris.

Rifkin, J., Chemla, F. and Chemla, P. (eds) (2012). *La troisième révolution industrielle : comment le pouvoir latéral va transformer l'énergie, l'économie et le monde*. Les liens qui libèrent, Paris.

Roger, F. (2010). Le marché des “emplois vert”, WWW Document [Online]. Available at: https://www.studyrama.com/formations/specialites/environnement-developpement-durable/le-marche-des-emplois-vert-45669 (accessed August 20, 2019).

Saini, P. and Shukla, K. (2016). Green recruitment: A new tool of cost cutting (conceptual study). *Int. J. Sci. Innov. Res.*, 4, pp. 195–198.

Sauvy, A. and Hirsch, A. (1980). *La machine et le chômage : le progrès technique et l’emploi*. Dunod, Paris.

Schumpter, J.A. (1943). *Capitalism, Socialism and Democracy*. George Allen & Unwin, London.

Schwab, K. (2017). *La quatrième révolution industrielle*. Dunod, Paris.

Stahel, W. (2016). Circular economy. *Nature*, 531, pp. 435–438 [Online]. Available at: https://doi.org/10.1038/531435a.

Stringer, L. (2010). *The Green Workplace: Sustainable Strategies that Benefit Employees, the Environment, and the Bottom Line*. Palgrave Macmillan, New York.

Tchernia, J.-F. (2005). Les jeunes Européens, leur rapport au travail. In *Les jeunes Européens et leurs valeurs : Europe occidentale, Europe centrale et orientale*, Galland, O., Roudet, B. (eds). La Découverte, Paris.

Waaub, J.-P. (1991). Croissance économique et développement durable : vers un nouveau paradigme du développement. In *Environnement et développement : questions éthiques et problèmes socio-politiques,* Prades, J.A., Vaillancourt, J.-G., and Tessier, R. (eds). Fides, Montreal.

Wehrmeyer, W. (ed.) (1996). *Greening People: Human Resources and Environmental Management*. Greenleaf, Sheffield.

Wijkman, A. and Kristian, S. (2016). L’économie circulaire et ses bénéfices sociétaux – des avancées réelles pour l’emploi et le climat dans une économie basée sur les énergies renouvelables et l’efficacité des Ressources, Report. Club of Rome.

World Economic Forum (2016). The future of jobs – Employment, skills and workforce strategy for the Fourth Industrial Revolution. Global Challenge Insight Report, World Economic Forum, Davos.

Zoogah, D. (2011). The dynamics of green HRM behaviors: A cognitive social information processing approach. *Z. Fuer Pers. Ger. J. Res. Hum. Resour. Manag.*, 25, pp. 117–139 [Online]. Available at: https://doi.org/10.2307/23279429.

4

The Role of Education in the Ecological and Circular Transition: Current Situation and Prospects

4.1. Introduction

Ecological transition, circular economy, industrial ecology and functional economy do not only question companies and organizations about their internal management methods and their relations with economic and social partners (Lazzeri *et al.* 2017). The emergence of new skills, even new professions, also calls on the academic world to fully mobilize and play its part in preparing future generations for the challenges of the 21st Century. Are higher education and research institutions and companies in the territory prepared to respond jointly to this new challenge? Should we create new training courses? Integrate this issue into existing training courses? Can we respond to companies' requests with lifelong learning actions? To answer these questions, we will rely on:

– the comments and debates gathered during the meetings of the French *Institut National de l'Economie Circulaire* in 2019, the round table "*Les besoins en compétences de l'économie circulaire*" (The skill requirements in the circular economy) which was held as part of the Aix-Marseille University Entreprise Week on November 4, 2015[1];

Chapter written by Dominique BONET FERNANDEZ.

1 This round table, which was part of the "Training, Professions and Territory" theme of the AMU/Enterprise 2015 Week, was organized by the Sustainable Development Department of Aix-Marseille University, in collaboration with the *Institut Régional de Formation à l'Environnement et au Développement Durable* (IRFEDD 2). It was summarized by B. Faye

– the latest legislative proposals on Education for the Circular Economy in September 2019[2];

– as well as the proposal of the UVED, *Université Virtuelle pour l'Environnement et le Développement durable* (Virtual University for the Environment and Sustainable Development).

This chapter presents the findings and challenges for the academic world, the possible responses in terms of training in ecological transition and, more particularly, in the circular economy, the solutions provided by UVED and, in conclusion and perspectives, the proposal for a circular economy law, education section, of 2019.

Since the beginning of the 2019–2020 academic year, students and stakeholders in higher education have been making their voices heard to ask the government and institutions for concrete measures to address the climate emergency[3]. One of these measures more particularly concerns the training of all students in circular economy and sustainable management. The Shift Project Report[4] published in spring 2019 highlighted the delay in French higher education in taking ecological issues in training into account. The forums published since then have highlighted this need and demand for higher education transformation on the part of students and higher education stakeholders. Faced with this multitude of challenges, our focus is on training in the circular economy, which we would call an operational approach to the implementation of sustainable development.

4.2. Findings and challenges for universities and companies

Sustainable development issues and climate-energy issues are still poorly taught, and *a fortiori* in post-high school training, which each year welcomes 6.5% of the "working-age" population (15–64 years old). Indeed, a recent study conducted by the Shift Project[5] with 34 higher education institutions shows that 76% of courses do not offer any courses addressing climate-energy issues to their students. In the remaining 24% of courses, less than half (11%) offer at least one mandatory course for students addressing the subject, and for good reason: the French Education Code,

et al. (2017) in the book *Economie Circulaire et Territoires*. Presses Universitaires de Provence, Aix-en-Provence.

2 http://www.assemblee-nationale.fr/15/propositions/pion2263.asp.

3 https://refedd.org/loi-former-enjeux-ecologiques/.

4 Mobiliser l'enseignement supérieur pour le climat. Report, The Shift Project, May 2019. Available at: https://theshiftproject.org/article/nouveau-rapport-mobiliser-superieur-climat/ (accessed: June 3, 2019).

5 https://theshiftproject.org/wp-content/uploads/2019/03/2019-03-25_Rapport_Mobiliser-lenseignement-sup%C3%A9rieur-pour-le-climat_The-Shift-Project.pdf.

like the other regulatory texts concerning the missions of higher education, does not mention the responsibility of institutions in this field.

The student mobilizations of recent months and the signatories of the "Student Manifesto for an Ecological Awakening"[6] not only call on the public authorities to assume their responsibilities, but also demand to be trained in these issues, in order to prepare them to their professional careers[7].

4.2.1. *Multidisciplinary challenges, requiring a review of training courses*

Emerging society issues concern major challenges: climate change, resource depletion, poverty, world hunger, etc., all of which are sustainable development goals that must be addressed in an interdisciplinary way in all training courses. The circular economy makes it possible to address the question of sustainable development using operational and operating business models, and can be broken down into a series of practical training courses. Faced with the difficulty of training for the "new topic" with its share of uncertainty and insufficient modeling, the challenge for the education system is to revisit its training practices.

The training courses are insufficiently multidisciplinary. Admittedly, there are training courses in the environment, sustainable development, public health, mediation (cultural), CSR (Corporate Social Responsibility) and SSE (Social and Solidarity Economy), but the link between the Human and Social Sciences and the so-called "hard" sciences is unsatisfactory (Faye *et al.* 2017). Academics from different sectors do not know each other well, and often do not speak the same language. Thus, the first interdisciplinary meeting generally consists of jointly defining terms such as ecosystem, ecology and sociological approach, each associating with these words a vision inherited from their own training.

4.2.2. *The companies and universities concerned*

Companies and professionals, within their respective structures, are also an important factor in individual development, hence the fundamental role of training as a potential driver for the emergence of the circular economy at the societal level, and beyond sustainable development. It should be noted that we are also talking about working conditions and the issue of quality of life at work. However, there is a great heterogeneity among companies, between lack of knowledge, skepticism and commitment to sustainable development.

---

6 https://pour-un-reveil-ecologique.fr/.

7 http://www.assemblee-nationale.fr/15/pdf/propositions/pion2263.pdf.

Finally, there is the question of access to information, valuation and modeling. Faced with this intersectoral observation, the increase in skills and, above all, the greater versatility required of professionals are leading to a change in the expression of skills. They also, and above all, encourage non-technical and transversal situations and skills: the ability to fit into a collective framework and cooperate, the ability to adapt, critical sense, autonomy, etc.

In terms of training, this requires a transversal adaptation rather than a modular approach (Faye *et al.* 2017). The holistic approach thus becomes an essential criterion in the development of a training offer. This also implies giving meaning to the profession concerned and its developments, hence the importance of raising awareness for a common culture of the circular economy and sustainable development and its implications in the professional sector or profession concerned.

4.3. How to meet the skills and training needs of the circular economy?

4.3.1. *From Green Plan to education for sustainable development*

For Jeziorski *et al.* (2015), university education is undergoing rapid changes in the context of education for sustainable development. Relayed by the European authorities, then the States, the requirements towards universities have gradually increased. In France, since the beginning of the 2009–2010 academic year, higher education institutions have been developing a "Green Plan" for campuses[8].

> The Green Plan enables higher education institutions to integrate sustainable development into their operations and teaching. Article 55 of the Grenelle 1 law stipulates that "higher education institutions will draw up a 'Green Plan' for campuses for the 2009 academic year. Universities and *grandes écoles* may apply for labelling on the basis of sustainable development criteria…".

Box 4.1. *The Green Plan*

But higher education institutions must lead the way through education for sustainable development. Within this framework of the institutional integration of education for sustainable development, university training is therefore undergoing rapid change, but it is confronted with or based on the specific characteristics of each country, which may be institutional, financial or cultural. Wals and Blewitt (2010) provide an interesting analysis on the implementation of sustainable development in higher education, and conclude that institutions are entering a "third

8 http://www.campusresponsables.com/sites/default/files/ressources/lecanevas_plan_vert_-_enseignement_superieur.pdf.

wave" (after the environmental wave and the green campus wave) that focuses on teaching and learning about sustainable development. For these authors, should environmental education and education for sustainable development change student behavior or should they focus on improving the environment? Or on capacity-building and critical thinking?

4.3.2. *A wide range of skills*

In terms of skills, the challenge is to train people in the ability to fit into a holistic system, in cooperation, in interdisciplinarity, in adapting to novelty and uncertainty, in evaluating and transmitting the intangible. Recently, the "*Groupe technique sur les compétences* (GTC) *en DD*", led by the *Conférence des Grandes écoles* and the *Conférence des Présidents d'Université*, proposed a new version of a reference framework on competencies for sustainable development in higher education. This reference framework distinguishes five meta-skills characteristic of education for sustainable development:

– collective skills;

– forward-looking vision;

– systemic vision;

– competence in terms of responsibility and ethics;

– changes (support, follow-up, initiation, etc.).

It is necessary to integrate into existing training courses an awareness of sustainable development, the circular economy, the economy of functionality, etc. through specific modules, or by integrating these concepts into the training in the form of a guiding principle. When developing the training offer, it is necessary to adopt a new approach based on trades and skills rather than diplomas. MOOCs, badges and accreditations are moving in this direction.

4.3.3. *Towards new practices*

In a constantly changing society, the time devoted to experimentation is not always appropriate. Nevertheless, it would be advisable to enhance the processes of trial and error, experimentation and action to facilitate understanding of the issues at stake. This requires not only time, trust and dialogue on the part of the partners, but also adapted pedagogical methods, including the experimentation of active and participatory pedagogies (serious games, problem-based learning, etc.). Awareness of the issues and debate should make it possible to meet current challenges. Engaging stakeholders in dialogue, questioning representations, tracking the implicit, disrupting routines – all actions that will move the debate forward. Thus, the importance of a shared problematization associated with the challenge of democratic debate can be fully expressed (Faye *et al.* 2017).

For their part, companies do not always need to create new professions but must consider a different use and exercise of the profession. Companies and their senior executives are facing ever more pressing challenges. Sustainable management and business globalization require skills and knowledge on the social and environmental challenges and impacts of economic activity. With digitalization as a backdrop, artificial intelligence (AI) offers potential for measuring and reducing the social and environmental impacts of managerial decision-making[9]. The training of managers and staff throughout life is therefore a fundamental challenge for a successful transition and responds to SDGs 4 and 10 (see Figure 4.1) which emphasizes that quality education is the foundation for improving people's lives and sustainable development.

Figure 4.1. *Sustainable development goals. Source: https://www.un.org/sustainabledevelopment/fr/2015/09/26/onu-appelle-contributions-secteur-prive/*

On the initial training side, even if students generally lack a general knowledge of sustainable development, nothing is lost. As a first step, the Sustainability

9 https://business.lesechos.fr/directions-ressources-humaines/management/formation-elearning/0601636873742-comment-se-former-sans-cesse-aux-enjeux-du-monde-331318.php.
10 https://www.un.org/sustainabledevelopment/fr/education/.

Literacy Test[11], which allows students to assess their general culture in terms of sustainable development, could help to lay the first bricks.

> Sustainability literacy is a term that covers all knowledge, skills and mentalities that enable everyone to commit themselves to building a sustainable future and to make informed and relevant decisions in this regard. Simple to use, adaptable and deployed worldwide, the "Test", created by Sulitest, is a valuable tool to help organizations and individuals contribute to the Sustainable Development Goals (SDGs) (see Figure 4.1) and to participate in a global movement enabling everyone to understand, be able and willing to commit themselves to the collective construction of a sustainable future and actively contribute to the Global Agenda.

Box 4.2. *Sustainable literacy*

Suggestopedia, a learning method, could be applied, to promote the learning of new concepts that use sustainable development and the circular economy, without making it a discipline. Finally, to develop awareness, to move from a state of interdependence to a state of cooperation. Lifelong learning trainees have learned to distance their experience, while students have a sharper capacity for learning and questioning. Hence, the need, in training courses, to mix initial and continuing training audiences. In view of the development of new technologies, it is also possible to combine SD training and digital training to make SD more concrete, for example, through serious games. All professions are concerned by what constitutes a societal revolution: learners must integrate these elements to prepare for tomorrow's world and become agents of change themselves.

In a non-exhaustive way, avenues are discussed to prepare students for their integration into the professional world and thus face the challenges that await them in their professional life:

– prepare for entrepreneurship;

– encourage group evaluation;

– develop interactive pedagogies;

– decompartmentalize the fields of training and employment: this is done in the knowledge of experience rather than in the knowledge acquired at university;

– value the importance and role of lifelong learning;

– not create new professions but consider a different use and exercise of the profession;

---

11 https://www.sulitest.org/en/index.html.

– propose an economic model adapted to the challenges; economically speaking, a balance must be found, thus transforming intangible capital into a value of profitability savings. And how to institutionalize it? Financial leverage is important.

4.4. An initiative provided by the French Virtual University for Environment and Sustainable Development

In a proactive approach, UVED has addressed these issues. This actor could be a catalyst for change by offering everyone simplified access to these new pedagogies around the circular economy and its components. According to UVED[12], it is essential that all university students, high school students, citizens and future economic actors are informed, sensitized and trained on these vital global issues. It is therefore essential to continue to make knowledge on environmental issues accessible to all and to help institutions quickly equip teachers to better train their students in environmental issues, whatever their curriculum.

To date, there is little teaching on climate and environmental issues in university courses. However, all disciplines are concerned by the ecological transition, with each discipline facing environmental and sustainable development issues. In this context of ecological and climatic emergency, and the growing mobilization of high school students, it is urgent to integrate these challenges into higher education training, all disciplines combined.

The lack of a strategy in higher education hinders the implementation of the necessary training. Faced with this need, schools are confronted with pedagogical, political and financial difficulties and obstacles: the lack of teacher training in these issues, the lack of human and financial resources, time, tools or even the lack of legitimacy to provide such training, the problem of the compartmentalization of disciplines, the saturation of teaching models, the rigidity of structures, etc. UVED plays an important role in supporting higher education and research institutions, teachers and teacher–researchers by providing them with advice, pedagogical content and open access tools for training and to train their students in environmental issues and ecological transition and to use as a support or complement to their existing teaching. These digital educational resources and e-learning courses (MOOCs, thematic courses, etc.) are produced by scientists and validated pedagogically and technically, thus guaranteeing their quality.

The academic world therefore has no choice but to train for this change of model. Educating young people in the society of tomorrow means preparing them

12 *L'Université Virtuelle pour l'Environnement et le Développement Durable*, https://www.uved.fr/.

for the instability of the world, bringing them to a state of mind that allows them to balance a "hard" world with a "benevolent" world. This requires the development of new skills for trainers and a profound questioning of their profession.

In order to support this reform and accelerate access to new forms of education, a draft law is currently being debated in the National Assembly.

4.5. The proposed 2019 law on the generalization of education on issues related to the preservation of the environment and biological diversity and climate change within the framework of global boundaries

At a time when young people seem to be taking the lead in mobilizing for the climate, a draft law aims to reflect, in legislation, the desire, expressed both by the students of the "Student Manifesto for an Ecological Awakening" and by the "Call of Teachers and Principals", that higher education prepares all students for the vital challenges of the 21st Century. This is the proposal for a law on the generalization of education on issues related to the preservation of the environment and biological diversity and climate change within the framework of global boundaries.

In this proposal[13], according to the scientific reports of the IPCC[14] and IPBES[15], the climatic and ecological urgency imposes rapid and profound changes on our model of civilization. Facing the challenges of leaving fossil fuels, preserving biodiversity, resources and respecting global boundaries requires a population that is not only aware of them, but also informed and trained – especially as future elected representatives, readers, parents and professionals in all sectors (from health to industry, culture and the civil service). It is not only a question of preparing for so-called "green" jobs because, to varying degrees, all functions are concerned.

The success of the call "to train all higher education students in climate and ecological issues", published in July 2019, which has exceeded 7,400 signatories, including 80 heads of institutions, shows that higher education takes the measure of the issue, and is ready to act within the framework of their autonomy, with the support and recognition of the State. The Prime Minister himself recently stated that "ecological transition is not the sole responsibility of the Minister in charge of this transition, but the responsibility of all members of the Government, who must place their action within this simple but very ambitious objective". Higher education has a role to play. In October 2019, 80 deputies from all sides presented a bill to the French Parliament aimed at ensuring that "higher education prepares all students for

13 http://www.assemblee-nationale.fr/15/propositions/pion2263.asp.

14 https://www.ipcc.ch/report/srccl/.

15 https://www.ipbes.net/news/ipbes-global-assessment-summary-policymakers-pdf.

the vital challenges of the 21st Century", by generalizing "the teaching of issues related to the preservation of the environment and biological diversity and climate change". Through 20 articles, this proposed law aims, in particular, to amend the Education Code and other regulatory texts to include the responsibility of institutions in terms of teaching issues related to climate change and the circular economy.

4.6. Legislative proposals to be closely monitored

It is up to legislators to provide the impetus and define the framework for recognizing the missions entrusted to higher education in the transmission of knowledge about climate and ecological issues and the circular economy, which we describe as a technique for operationalizing the transition.

The proposal to amend the Circular Economy Law 2019, put forward by the National Institute of Circular Economy, recommends the following, in Measure 7, in favor of education[16].

> Training in the circular economy from an early age. Levers: Educated and trained in the challenges of the circular economy, aware that resources and raw materials must be used sparingly and efficiently, the next generation will be better able to collectively organize to find solutions to the depletion of natural resources and climate change. To raise awareness and provide training on circular economy issues, the Institute offers: 1) teaching modules to raise awareness and awaken children to the challenges of the circular economy from primary school onwards, 2) operational and academic training modules on the various aspects of the circular economy in all secondary and higher education courses and 3) vocational training to prepare them for new occupations linked to the challenges of scarcity of resources.

Box 4.3. *Proposal to amend the Circular Economy Law 2019*

The resulting proposal for Law 2019 proposes that higher education institutions systematically integrate the teaching of environmental preservation issues within the framework of global boundaries into their courses so that all students are trained in them, regardless of the course chosen. The circular economy appears as a filigree of these planned educational themes.

16 https://docplayer.fr/amp/139978579-Les-10-de-l-inec-propositions-pour-la-loi-economie-circulaire.html.

Article 1 includes, in the general principles of education, the teaching of issues related to the preservation of the environment and biological diversity and climate change within the framework of global boundaries.

Article 2 integrates compulsory and interdisciplinary education of these issues into the missions of higher education, the national strategy for higher education and the responsibilities of institutions, in an interdisciplinary approach, emphasizing the need to build on recognized scientific findings, and to highlight the implications for students' future careers.

Article 3 sets out this requirement in order to ensure that it is also present in the planning schemes and regional schemes for higher education, research and innovation, regional plans and the map of higher education.

Article 4 stipulates that the objectives of the regional training and vocational guidance development plan contracts include, as one of their objectives, that the challenges related to the preservation of the environment and biological diversity and climate change within the framework of global limits are effectively taught in each vocational training course, as well as their impacts on each vocational field.

Article 5 extends the subject to all public schools and educational establishments.

Article 6 specifies that students are to be taught both the scientific and societal fundamentals of these issues, as well as the implications for the different professional fields.

Article 7 mandates the Academic Council to ensure that the above-mentioned issues are properly taken into account in research programs, in order to promote their teaching by teacher–researchers.

Article 8 makes it possible to create a common service for universities to promote interdisciplinary teaching of the above-mentioned issues.

Article 9 provides that multi-year establishment contracts shall consider national objectives, in particular those of the National Low Carbon Strategy, and that their activities shall contribute to achieving them.

Article 10 provides that educational projects and research programs shall consider national objectives, in particular those of the National Low Carbon Strategy, and that their activities shall contribute to achieving them.

Article 11 specifies that higher education must consider national objectives, in particular those of the National Low Carbon Strategy.

Article 12 specifies that the institution's general policy must consider national objectives, in particular those of the National Low Carbon Strategy, and provide that the institution's activities contribute to achieving them.

Article 13 provides that the university's board of directors shall also elect a vice-president responsible for considering ecological issues, and, in particular, for teaching all students about them.

Article 14 provides that representatives of approved environmental protection associations shall sit on the board.

Article 15 provides that the higher national institutes of professorship and education shall provide training in issues related to the preservation of the environment and biological diversity and climate change within the framework of global boundaries.

Article 16 coordinates in the Energy Transition for Green Growth Act to ensure that the training of the French population is consistent with national energy transition objectives.

Article 17 replaces the notion of training in sustainable development issues with that of issues related to the preservation of the environment and biological diversity and climate change within the framework of global limits, in Act No. 2009-967 of August 3, 2009, on the implementation of the *Grenelle de l'environnement.*

Article 18 coordinates, in Act No. 2009-967 of August 3, 2009, on the implementation of the *Grenelle de l'environnement*, in order to specify the expectations of higher education institutions in terms of training in climate, energy and environmental issues.

Article 19 specifies the orientations of research in terms of considering environmental issues, and provides for the alignment of the national research strategy with the objectives of the National Low Carbon Strategy.

Article 20 provides that civil servants must be able to be trained in environmental conservation issues.

4.7. Conclusion

Faced with the climate emergency, more and more voices are being raised and advocating for the generalization of training for sustainable development and its operational applications such as CSR and the circular economy, all sectors combined, in an interdisciplinary approach. The circular economy offers a field of multidisciplinary knowledge and skills, ranging from the ecodesign of the product to its responsible production, distribution and consumption, reparability, reuse and recycling, in an industrial ecology approach. While the current state of play shows a delay in the educational world taking these issues into account, various initiatives, which we have reported on in this chapter, indicate that reform is under way.

A sufficiently powerful learning process is required to achieve transitions that require a change in values, corporate culture, lifestyle and, ultimately, a complete overhaul of the system. But what are these skills and how can they be developed? This research challenge can be addressed in a sustainable development context. In addition to the technical and technological skills required, a set of soft skills must be highlighted. Our society is not trained to cooperate, so it is a question of creating capacities to cooperate and developing the skills associated with cooperation: respect, welcoming diversity, openness to others, accompanying change (Faye *et al.* 2017) and compassion. The new educational model requires increased skills in cooperation and adaptability. Cooperation and trust are new skills needed in today's markets. Companies are looking for new ways to build customer relationships. Consumers now buy functions rather than objects. The required social learning is at the crossroads of informal, non-formal and formal education (Wals 2011). Such a context also includes community social learning and lifelong learning, but always in relation to educational institutions and organizations, within a stimulating regulatory framework and with an appropriate territorial and local base.

4.8. References

Faye, B., Domeizel, M., Delannoy E. and Richaud I. (eds) (2017). Comment l'université peut-elle répondre aux besoins en compétences de l'économie circulaire ? In *Economie circulaire & territoires*, Lazzeri, Y., Bonet Fernandez, D. and Domeizel, M. (eds). Presses Universitaires de Provence, Aix-en-Provence..

Jeziorski, A., Legardez, A., Floro, M. and Domeizel, M. (eds) (2015). Quelles compétences pour l'éducation au développement durable dans l'enseignement supérieur ? Une étude exploratoire à l'Université d'Aix-Marseille. *Colloque international francophone "Former au monde de demain"*, Clermont-Ferrand, France, 2–3 April.

Lazzeri, Y., Bonet Fernandez, D. and Domeizel, M. (eds) (2017). *Economie circulaire & territoires*, Presses Universitaires de Provence, Aix-en-Provence.

Wals, A.E.J. (2011). Learning our way to sustainability. *Journal of Education for Sustainable Development*, 5(2), pp. 177–186.

Wals, A.E.J., and Blewitt, J. (2010). Third wave sustainability in higher education: Some (inter)national trends and developments. In *Green Infusions: Embedding Sustainability across the Higher Education Curriculum*, Jones, P., Selby, D. and Sterling, S. (eds). Earthscan, London.

5

Boosting the Circular Economy through Proximity: The New Competences of Local Authorities

The deployment of the circular economy on a territory depends closely on the coordination efforts of local actors. As stated in the 2014 report of the CGEDD, the aim of the circular economy is to promote, "cooperation between economic actors at the relevant territorial level". However, defining the proper or relevant territorial scale has been consistently addressed in academic research on sustainable development actors. The organization of flows in a circular economic process requires an understanding of the types of actors and the proximity that brings them together. This is one of the topics addressed by the theoretical framework of the academic school of proximity, which for nearly 25 years has been proposing to rethink the effects of distance on interactions between actors through a non-geographical prism. Among the local actors, local authorities play a key if not essential role. Indeed, in France, local authorities and, in particular, municipalities (individually or in an urban community) have acquired a historical expertise in the collection and recycling of waste; competences that have been reinforced by the transfer of responsibility carried out by the State in the course of successive territorial reforms and, in particular, those of Act III of decentralization (2013–2015). However, in recent years, municipalities have moved from a logic of waste disposal, in the context of maintaining hygiene and public health, to a global action of circularization of the local economy, whether through the reuse of waste of all kinds, a public procurement policy taking into account environmental impacts, or the development of associative initiatives, particularly aimed at developing new consumption practices among the population (repair cafés, disco soup events, etc.). The municipalities have thus gradually become involved in the development of local

Chapter written by Karine FABRE and Alexis POKROVSKY.

initiatives for a circular economy, by providing partners with their know-how and skills. In fact, these initiatives have a positive impact on their territory, for example: the creation of activities that cannot be relocated or a response to citizens' concerns. In addition, recent developments in territorial organization in France, for example: the reinforced grouping of municipalities within intermunicipal organizations; the new status of metropolitan areas introduced by the law on the modernization of territorial public action and the affirmation of metropolitan areas (MAPTAM) of January 27, 2014; as well as international cooperation projects, have created a favorable environment for sharing experience and pooling circular economy projects, beyond the municipal framework. This chapter thus proposes to present, in light of recent legislative advances, the growing role that local and regional authorities are expected to play in the development of circular economy initiatives, whether in partnership, or in support of the various actors. This chapter is structured into three sections: first, we will present some lessons on local scale measurement supported by academic work based on the school of proximity. Second, we will recall the respective competences of the different territorial levels in the field of the circular economy. Finally, in the third section, we will focus more specifically on the different levels of action available to public actors, highlighting them with examples.

The work and debates on the draft low-waste law, during the 2019 parliamentary session, made it possible to recall the importance of positioning local public governance structures in the deployment of circular economy initiatives; indeed, everyone should be able to acquire a better knowledge of public support mechanisms at the so-called "local" level and the multiplier effect that local actors (elected officials, administration, associations) can bring to the efforts of the sector as a whole.

5.1. What scale for proximity?

Circular economy circuits are based on flow loops organized around appropriate territorial scales: the positive effects obtained in the treatment or reprocessing of materials shall not lead to additional CO_2 and pollutant emissions via long-distance transport. The proximity scale appears as the central notion, around which the organization of flow loops is based.

Following this line of reasoning, all legislative texts have consistently stressed the need for proximity as a principle of action in the field of the circular economy within a territory. However, the definition of this proximity in the existing texts remains vague, and above all, not very operational as far as the question of the adequate scale of application is concerned: city, region, nation, world?

The academic school of proximity has precisely developed a theoretical corpus demonstrating that there is no canonical measure of proximity but, on the contrary, a typological diversity of proximities. This theoretical current sheds light on the complex way in which geographical, organized, social, cognitive and institutional proximities frame the interactions between actors – whether private or public – essentially questioning the way in which a territory is structured through these proximity dynamics (Torre and Talbot 2018). In terms of ecology, the authors of this trend, very early on, examined the role that proximity plays in environmental negotiation procedures (Torre and Beuret 2012). They invited us to go beyond the horizon of the administrative division, in which the role of local authorities is regularly confined, in order to rethink our scales of action according to the various modalities of proximity.

Thus, research carried out on different waste treatment sectors identified that the effective territorial scale for waste treatment depends, first of all, on the organization of value chains (Durand *et al.* 2016): to the local territory, heavy and putrescible waste (here the municipality and intermunicipalities), to the national territory, hazardous waste management and to the international scale, high value added waste. The so-called extended producer responsibility or EPR channels are based on a scope essentially determined by the industrial logic of processing.

The spatial organization of industry resources, such as the concentration aimed at generating economies of scale, or the specialization of certain territories in high value added activities, may explain the reasons for this tiered approach to waste treatment on several geographical scales. Organized proximity and geographical proximity thus complement each other in designing a spatial organization of material flows.

An additional dimension of organized proximity completes this picture. Organized proximity, i.e. the "capacity that an organization provides to create interactions among its members" (Rallet and Torre 2004), mainly presents itself as a relational capacity that promotes coordination actions. It facilitates resolution of conflicts of environmental uses caused by imposed geographical proximity. Conflicts arise when actors with asymmetric access to space are constrained by geographical proximity and compete in the establishment of new forms of proximity (Torre and Zuindeau 2009). In local circular economy projects, some inhabitants may indeed fear environmental nuisances, generated at the local level. In this context, the organized proximity provides collective solutions to resolve all these conflicts. It is therefore easy to understand that in the above-mentioned case of the national location of hazardous waste treatment, in addition to an economic motivation, the tensions and conflicts linked to the exploitation and storage of hazardous waste on a local territory will press the actors (municipality, region, State) to resolve local conflicts through an organized solution, such as grouping treatment in specialized locations. However, as the authors (Torre and Zuindeau 2009) point out, organized proximity can in turn also generate new types of conflicts, particularly when it thwarts the shared values of local communities, pushing geographically close actors to engage in forms of resistance.

It is ultimately a complex set of proximity arrangements that should be considered here. Three major approaches are emerging in order to integrate the territorial scale into projects involving private and public actors, around the circular economy (Brullot *et al.* 2017): sharing a common vision and information base; affirming the role of organizational and institutional innovations as well as technical innovations; and finally, providing for an evaluation of projects to measure the actual impact on the territory. Mention should be made here of the complementary role of institutional proximity, which is based on the sharing of common rules and codes, and facilitates anticipation and common behavior among stakeholders (Zimmermann 2008). To this end, it is essential to know each other's actions and roles in the service of a circular economy project; the following therefore aims to shed light on the essential functions of territorial public actors in this field, as well as their levers for action.

5.2. Understanding the territorial levels: sharing of competences in the context of waste management

In order to better understand the tiering of the different competences shared between the levels of local authorities, we propose to illustrate the role of each level in the case of waste management. Indeed, waste management, if it does not summarize all the actions of the circular economy, represents the historical link that made it possible to build the competence of public actors in the organization of material flows at the scale of a territory. We will first look at how communities have built their competence through a historical reminder. Then, we will discuss the current role of each territorial layer in waste management.

5.2.1. *Historical expertise in waste management acquired by municipalities*

Until the 19th Century, the concept of hygiene was not well developed. It was in 1870 that Louis Pasteur highlighted the link between hygiene and health. It is then in 1884 that Eugène Poubelle, Préfet de Paris, ordered the deposit of waste in special containers equipped with a lid (subsequently called "*poubelles*" in French) outside of houses, so that the waste would no longer be scattered in the street and could be collected by the municipal services. This practice was then followed by many cities but remained underdeveloped in rural areas. As waste management was not subject to national regulations, each municipality organized itself as it saw fit.

It was almost a century later, on July 15, 1975, that the first major law on waste management was enacted, requiring each municipality to collect and dispose of household waste. Communities organized themselves, some choosing incineration, others landfill. The 1980s, due to changes in lifestyle and consumption, saw waste quantities increasing to double in 40 years. Most of the waste was then sent to landfill, misusing an enormous amount of raw materials and raising problems of soil pollution.

The second structuring law on waste management was enacted on July 13, 1992. The perspective was changing. The law provided for the prevention and reduction of waste at source, selective sorting (mandatory from that date), recovery through recycling and reuse. The law planned for the gradual abandonment of landfills for classified installations and imposed the principle of extended producer responsibility with a financial contribution, from which eco-organizations, including eco-packaging, were created. Local authorities must now organize themselves to eliminate illegal dumping and recover waste through recycling, composting or clean incineration.

To this end, the Mayor holds (article L2212-2 of the General Code of Territorial Communities) the power of municipal police in order, "to ensure good order, public safety and health". It is in this capacity that the Mayor intervenes in terms of cleaning public spaces, removing bulky items, safety and convenience of passage in the streets. The General Code of Local Authorities also establishes the responsibility of municipalities for the disposal of household waste (Article L2224-13). This competence and responsibility must be transferred to the public intermunicipal cooperation body to which the municipality belongs.

5.2.2. *Territorial tiering of competences*

5.2.2.1. *National scale*

At the national scale, waste management planning has been mandatory in France since Act No. 92-646 of July 13, 1992 on waste disposal. In response to the requirements of the European Waste Framework Directive 2008/98/EC, the scope of planning has been extended. As a result, a national waste prevention plan (PNPD) has been approved for the period 2014–2020. In addition, a national waste management plan is being developed (public consultations in April and May 2019 followed by a provisional decree). This plan is above all an educational tool for presenting national ambitions in a single document. It should be noted that, because of their degree of harmfulness or particularities of management, some categories of waste listed by decree in State Council may require specific national prevention and management plans (Article L541-11-1 of the Environment Code).

5.2.2.2. *Regional prerogatives*

Act No. 2015-991 of August 8, 2015 on the new territorial organization of the French Republic (*nouvelle organisation territoriales de la République*, NOTRe), which marked the completion of Act III of decentralization, strengthened the powers of the regions, and those of the groups of municipalities, in the form of EPCI (*établissement public de coopération intercommunale*, "public establishment for intermunicipal cooperation") in the field of waste management.

The regions are now responsible for drawing up regional waste prevention and management plans with an extended scope.

Since 2004, France has committed to the development of a first five-year national plan of actions to prevent the production of waste, based on the premise that, "the best waste is the one that is not produced". As a European pioneer in this field, France was one of the first countries to move from a logic of optimizing the treatment of waste produced to the implementation of concrete actions, aimed at reducing waste at source (i.e. during the manufacturing, transportation and distribution stages of the product) and at the consumer level (responsible purchase and use, repair, reuse).

The 2014–2020 Waste Reduction and Recovery Plan takes a further step by addressing all the levers of action associated with prevention through 54 tangible, more transversal actions covering more waste flows. The objective set by this plan was to reduce household and similar waste by 7% by 2020 and to at least stabilize the production of waste from economic activities and construction.

Box 5.1. *France as a European pioneer in waste prevention*

The law establishes a Regional Waste Prevention and Management Plan (*Plan régional de prévention et gestion des déchets*, PRPGD) under the competence of the Regional Council and intends to promote the circular economy. These regional plans take over from the national plans that have followed one another since 2004. The objective is to produce a shared territorial diagnosis of waste management and anticipate its evolution over the next 6 to 12 years. These plans also make it possible to coordinate all the actions undertaken by both public authorities and private bodies to ensure the achievement of the objectives of the national waste prevention and management policy, and to contribute to the transition to a circular economy. They specify the prevention, recycling and recovery objectives to be reached, as well as the actions, services and equipment to be deployed. They have a legal impact because decisions taken in the field of waste by public legal entities and their concessionaires must be compatible with these plans. Beyond these legal impacts, regional plans help in defining, leading and managing local waste prevention and management policies.

The plans are subjected to annual monitoring and a more precise evaluation (every six years) likely to lead to their revision, within the framework of broad consultation via a consultative commission for the elaboration and monitoring of each plan. They are also submitted to public inquiry before adoption by the Regional Council, and the results of the evaluation of their implementation must be made public.

5.2.2.3. *Strengthening the role of intermunicipal authorities*

A major legal change impacts the competence in the collection and transfer of household waste. While previously the competence of intermunicipal authorities was only optional, the law now transfers it to them automatically.

In practice, for reasons of more relevant territorial scale, economies of scale and cost rationalization, this service had, for several decades, often been transferred to intermunicipal authorities.

Local authorities collect and treat waste on a direct basis (with their own staff and by maintaining the management of the service), or by calling on a private company, usually through a contract or a public service delegation, after a call for competition in accordance with the Public Procurement Code or the General Code of Local Authorities.

The public waste service is financed by a tax or charge that can be defined in such a way as to encourage the reduction and sorting of waste. Indeed, the cost of this service is important for communities. It had been estimated by ADEME in 2013 at more than 10 billion euros[1], with a per capita cost varying from 57 to 110 euros per year, depending on territorial characteristics.

> The term intermunicipality refers to the different forms of communal cooperation: municipal unions, communities of communes, communities of conurbations, urban communities, metropolitan areas and so on. These forms of cooperation are more or less integrated, ranging from intermunicipal management and sharing of some services (school transport, waste, etc.) to an intermunicipal project.
>
> Some intermunicipal entities are called own-tax entities because they generate their own financial resources through dedicated taxes (agglomeration communities,

1 Source: https://www.ademe.fr/expertises/dechets/elements-contexte/politique-vigueur/dossier/cadre-reglementaire/obligations-collectivites-cadre-service-public.

communities of municipalities, etc.). Intermunicipal associations without their own taxation (trade unions) benefit from direct budgetary contributions from their members. Public institutions for intermunicipal cooperation (*établissements publics de coopération intercommunal*, EPCI) have a deliberative body composed of elected representatives appointed by the municipalities (EPCI without own taxation) or directly by the voters during municipal elections. The deliberative body functions as a municipal council.

It should be noted that today more than half of the expenditure of local authorities comes from the local level, i.e. from municipalities and bodies with their own taxation (54% in 2016, against 36% in the departments and 15% in the regions). These figures highlight the important role that the local level has played in economic decisions at the territorial level. The purpose of this increase in local expenditure is to bring financial decisions closer to a "proximity" action (Delpech and Navarre 2018).

Box 5.2. *What is meant by intermunicipality?*

5.3. The levers for steering the circular economy

The management of the circular economy by local authorities takes a variety of forms, and we have discussed the role that each territorial level is called upon to play in the orchestration of actions, with a view to deployment.

Thus, the region has acquired a pivotal role in coordinating actions between the multiple actors in its territory with industrial ecology projects, whether they are companies seeking outlets for their production or developing a project with a circular economy purpose, or associations. A first avenue concerns the roll out of activity zones that could be considered as the ideal level for an effective combination between geographical and organized proximity (Georgeault 2017). A second approach consists of the creation of regional knowledge bases, in collaboration with partners such as ADEME, in order to promote initiatives on their territory and to share good practice. Regional segmentation remains a brake that limits the diffusion and the training effect, due, in particular, to the diversity of the solutions presented.

The municipal or intermunicipal level, through its active management of many flows on its territory, as we have illustrated above in the case of waste, is in direct contact with many operational decisions that can result in the increase of circular economy projects or initiatives.

However, if there is one area that brings together all the local actors, it is that of public procurement. The lever of public procurement is indeed a major tool for the development of actions at local level to promote the circular economy. Indeed, as seen in Box 5.2, more than half of public expenditure is channeled through the local level. In terms of total public procurement (i.e. including the State and local authorities), local and regional authorities are by far the main contractors and account for 58% of the total amount of public procurement, or more than 45 billion euros (Public Procurement Barometer: Results 2018 – ACDF). By virtue of its basis, this local expenditure would therefore make it possible to guide the future choices of the local economy towards sustainable development solutions. However, it seems to us that the mechanisms underlying it are insufficiently known and deserve to be developed.

5.3.1. *An increasingly environmentally oriented public procurement*

Public procurement is intended to apply to a wide range of products and services (collective catering services, administrative supplies and furniture, transport services, work contracts, road cleaning, public lighting leases, etc.). The specific rules of public procurement, compiled in the Public Procurement Code that came into force on April 1, 2019, apply to public actors operating both at the national (State) level and at infra levels of territorial responsibility (regions, departments, metropolitan areas, cities, etc.). These rules aim to guarantee the proper use of public funds by ensuring compliance with three main principles: freedom of access to public procurement, equal treatment of candidates, and transparency of procedures.

The French and European legal framework for public procurement has been constantly evolving over the past 15 years to promote the consideration of sustainable development and the circular economy in public procurement. Thus, in 2006, the regulations were amended to take environmental criteria into account, in particular with the introduction of the "best bidder" rule, which reduces the price criterion to one criterion among others (quality, social, environment, etc.). The criteria for choosing the best bid must be set out at the outset of the public contract in order to comply with the principles of equal treatment of candidates, whose bids will be judged exclusively on the basis of the objective criteria set out above, and transparency, since these criteria are known to all in advance. Until 20 years ago, the majority choice criterion was mainly financial. Traditional weightings consisted of assigning a weighting of 70% on the price score (lowest price bidder) and a maximum of 30% for technical value (quality of products and services, delivery times, after-sales service). From now on, it is not uncommon to see qualitative criteria equal to or more than the price. These qualitative criteria are not limited to a purely technical value of the offer but take into account the environmental and social impact aspects of the purchase.

It is with the European Directive 2014/24/EU that public purchasers were also able to refer explicitly to the lifecycle of a product in the definition of the subject matter of a contract and were entitled to introduce criteria relating to the circular economy into the technical specifications and performance conditions of a contract.

It then became possible to take into account all the costs generated by a product throughout its lifecycle and to encourage candidates offering products from the circular economy that tend to reduce the quantity of materials and energy consumed and, downstream, that limit the costs associated with the end of life of the products (reuse, repurposing, recycling).

Act No. 2015-992 of August 17, 2015 on the energy transition for green growth required the State and local authorities to achieve, by 2020, the conversion of at least 70% of the materials and waste produced on road construction or maintenance sites, for which they are the project owner.

However, these commitments have not been included in regional or departmental waste management plans and no details have been given on how to monitor and control compliance with these commitments. In addition, none of the above provisions are binding. There is thus a certain difficulty for public actors to grasp the legal tools and mechanisms intended to introduce sustainable development and the circular economy into public procurement in France.

5.3.2. *Some examples of public practices or policies: between responsible purchasing and waste reduction at source*

At the European level, there are interesting initiatives that could easily be transposed into France. For example, in Denmark, in 2012, a purchasing group including about 60 municipalities purchased administrative furniture made from 70% recycled or certified sustainable wood at a lower cost than the market price. In the Netherlands, in 2015, the city of Brummen launched a call for projects for the renovation of its city hall, as part of a full circular approach. During the purchasing process, price and environmental criteria had the same weight in relation to the final score, which enabled the city to build a building that was planned to last 20 years, with several elements made of wood (load-bearing structure, facade, floors), easy to dismantle and 95% of the materials that make it up can be reused in 20 years to build other buildings. All this was achieved at a lower cost than a traditional construction.

In France, projects are beginning to emerge. Thus, the City of Lyon has set up a web platform "*Achat Local Conso Rhône-Alpes*" in order to create a link between producers and buyers, as part of the public contracts related to school meals.

Since decree no. 2016-360 of March 25, 2016, buyers have finally had the opportunity to carry out, upstream of their market, a sourcing process that allows them to better identify the innovative companies in their territory in order to then adapt their specifications and facilitate access to their markets for these companies.

However, in practice, political constraints as well as the pace and working methods within public organizations do not favor the mobilization of this tool.

In order to substantially increase the percentage of environmental criteria integrated into public procurement, it is essential that guides, documents and training be put in place to support stakeholders and help them to change their daily practices. In addition, within local authorities, projects must be handled in a cross-functional manner with a constructive partnership between the division in charge of purchasing and the one dealing with environmental issues. A strong example in this regard is the Annecy Metropolitan Area Community, which has embarked on a proactive policy of buying less, but better, while preserving resources. To this end, services have been trained in the introduction of environmental clauses in contracts and specific tools have been proposed to them (calculation of the overall cost of using a product, thematic guides, etc.).

5.4. Conclusion

In conclusion, local authorities are proving to be an important player in the development of the circular economy in France. First of all, in terms of available resources. Indeed, territorial tiering offers a network and scales adapted to the scope of the different projects, or to the personality of the different actors (companies, large projects, associations, etc.), from the region, pilot of the organization of a global strategy, to intermunicipalities, capable of pooling local resources to achieve a critical mass necessary for the viability of many projects. Second, local authorities have established themselves as actors with skills: indeed, we have seen that the historical skills acquired by local authorities in waste treatment offer know-how on the management of material flows at the scale of a territory, both on a technical level and through mastery of local issues and an ability to resolve potential conflicts. Finally, the lever of public procurement is a powerful tool for initiating economically viable projects, given the amounts involved and the political will to move towards responsible public procurement.

5.5. References

CGEDD (2014). L'économie circulaire, état des lieux et perspectives. Report, November 2014.

Brullot, S., Junqua, G. and Zuindeau, B. (2017). Écologie industrielle et territoriale à l'heure de la transition écologique et sociale de l'économie. *Revue d'économie Régionale et Urbaine*, 5, pp. 771–796.

Delpech, C. and Navarre, F. (2018). Finances publiques locales et enjeux territoriaux. *Revue d'économie Financière*, 132(4), pp. 91–106.

Durand, M., Bahers, J. and Beraud, H. (2016). Vers une économie circulaire... de proximité ? Une spatialité à géométrie variable. *Déchets Sciences et Techniques*, 71, pp. 49–63.

Georgeault, L. and Lambert, F.-M. (2016). Logistique et écologie industrielle dans la construction d'une politique publique d'aménagement du territoire. In *Économie circulaire et territoires*, Bonet, D., Domeizel, M. and Lazzeri Y. (eds). PUP and PUAM, Aix-en-Provence.

Rallet, A. and Torre, A. (2004). Proximité et localisation. *Économie Rurale*, 280(1), pp. 25–41.

Torre A. and Beuret J.-E. (2012) *Proximités territoriales. Construire la gouvernance des territoires, entre conventions, conflits et concertations*. Economica, Paris.

Torre, A. and Talbot, D. (2018). Proximities: Return over 25 years of analysis. *Revue d'économie Régionale et Urbaine*, 5–6, pp. 917–936.

Torre, A. and Zuindeau, B. (2009). Les apports de l'économie de la proximité aux approches environnementales : inventaire et perspectives. *Natures Sciences Sociétés*, 17, pp. 349–360.

Zimmermann, J.-B. (2008). Le territoire dans l'analyse économique. Proximité géographique et proximité organisée. *Revue française de gestion*, 34(184), pp. 105–118.

PART 2

Circular Economy: A Few Tools and Approaches

6

Example of a Pioneering Approach to the Circular Economy: Cradle to Cradle

During the 20th Century, developed countries and then the major emerging countries became involved in mass consumer society. The productive model underlying this economic development is known as the linear model: extracting, processing and disposing of natural resources. It is accompanied by a constant and exponential increase in the materials and energies used.

What seemed to be an inexorable progress towards a society of abundance has turned against itself: raw materials are not available in infinite quantities and the pollution generated has irreversible repercussions on our health, biodiversity and climate. We are living with climate change as one of the greatest crises humanity faces. The call of 700 French climate scientists in September 2018 was very clear:

> Only immediate changes and short-term commitments, within the framework of clear and ambitious objectives by 2030, can enable us to meet the climate challenge[1].

In recent decades, the concept and term "sustainable development" has affirmed the need to reduce the impact of human activities: to pollute less, consume less energy and natural resources and recycle more. This is a first step, but not enough to address the issues. We need to change the perspective and purpose.

Chapter written by Anne DE BÉTHENCOURT

1 Réchauffement climatique: Nous en appelons aux décideurs politiques. *Libération,* September 7, 2018. Available at: https://www.liberation.fr/planete/2018/09/07/rechauffement-climatique-nous-en-appelons-aux-decideurs-politiques_1677176.

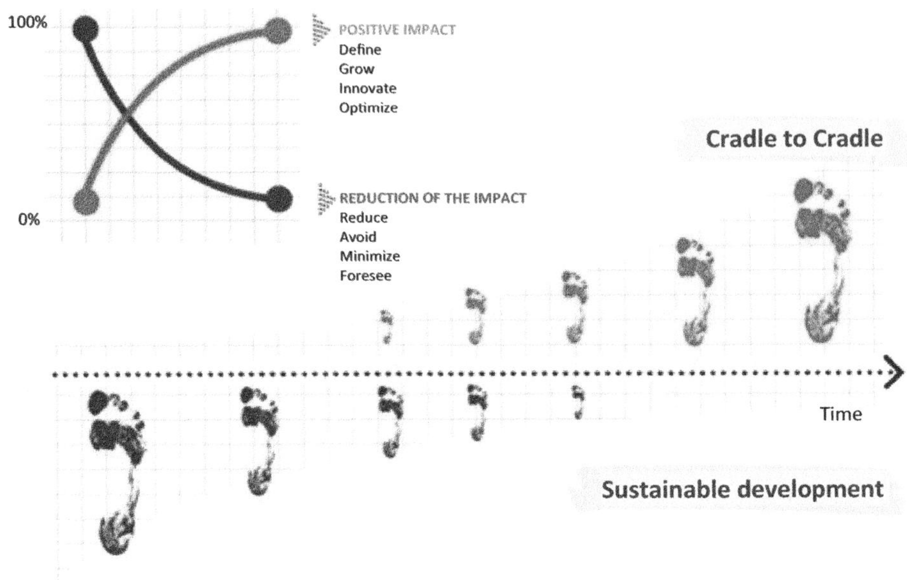

Figure 6.1. *Towards creating a positive impact rather than reducing the negative impact (source: Upcyclea based on MBDC[2]). For a color version of this figure, see iste.co.uk/delchet/circular.zip*

Changing paradigm means finding the origin of the word "economy" meaning "the one who runs the house" and aiming for "prosperity without growth", to use the words of the economist Tim Jackson (2010). The aim is to decouple economic growth from growth in natural resource extraction.

Can we reinvent forms of production and consumption in each sector of the economy that create a positive impact and not reduce negative impacts?

6.1. The origins of Cradle to Cradle

6.1.1. *Thus, Cradle to Cradle was born*

In the 1990s, German chemist Mickael Braungart and architect William McDonough created Cradle to Cradle (C2C), which is both a philosophy and methodology for creating a healthy world and eliminating the notion of waste. This approach is inspired by the way nature works, in which waste does not exist. The energy used is renewable, such as solar energy. Diversity is the norm. At the end of

2 https://www.upcyclea.com/cradle-to-cradle.

life, everything becomes a nutrient again. Everything is useful for something. No one would ever think of cleaning up a forest by removing all the dead leaves. When they fall to the ground and wither, they become, not waste, but nutrients for the living ecosystem that benefits from them. The loop is closed and natural balance is assured.

Cradle to Cradle therefore assumes that humanity can have a beneficial, positive and restorative impact on the environment. Thus, rather than aiming for the "less bad", Cradle to Cradle proposes that we aim for the "better".

6.1.2. *The founding principle: design differently and "upcycle" to create "positive loops"*

To avoid wasting resources and polluting, it is necessary to rethink the way resources are used and to design "material banks". This means designing products differently: we must be able to disassemble them, they must be repairable, non-toxic and their materials "almost infinitely" reusable. In this way, eco-designed products that reach the end of their useful life – and not the end of their life – will be processed. The value of the material thus preserved will return to a new production cycle, rather than end up in the dumpster. The intention here is to eliminate the notion of waste and preserve the value of resources. Creating positive loops and maintaining the value of the material is called upcycling. This is the fundamental principle of the Cradle to Cradle philosophy.

In concrete terms, engaging in a circular economy approach with a positive impact Cradle to Cradle methodology for the design of goods, for example, will therefore consist of selecting non-toxic components, choosing suppliers who respect human rights and employee health, promoting the use of renewable energy and ensuring that the water used in the production process is not contaminated. The materials used in different manufacturing processes will have to be able to either return to the ground (biological cycle) or be recycled within industries (technical cycle).

In the biological cycle, materials are broken down by microorganisms to form new nutrients. Biodegradable products are transformed into compost, thus forming a base of nutrients that can be used by new natural resources. All products that belong to this cycle are called "consumer products". In the C2C vision, clothing, footwear, tires or any other product that wears out during use must be designed with a view to returning to earth, only using substances that are compatible with the biological cycle. Usually, what we call "recycling" is more like "subcycling" or "down cycling" because it is often a source of value destruction. Indeed, the energy recovery of organic matter results in the loss of precious nutrients. In a closed and positive impact cycle, a product is eco-designed from 100% biodegradable components and can therefore return to the soil from which it is produced to be used as compost.

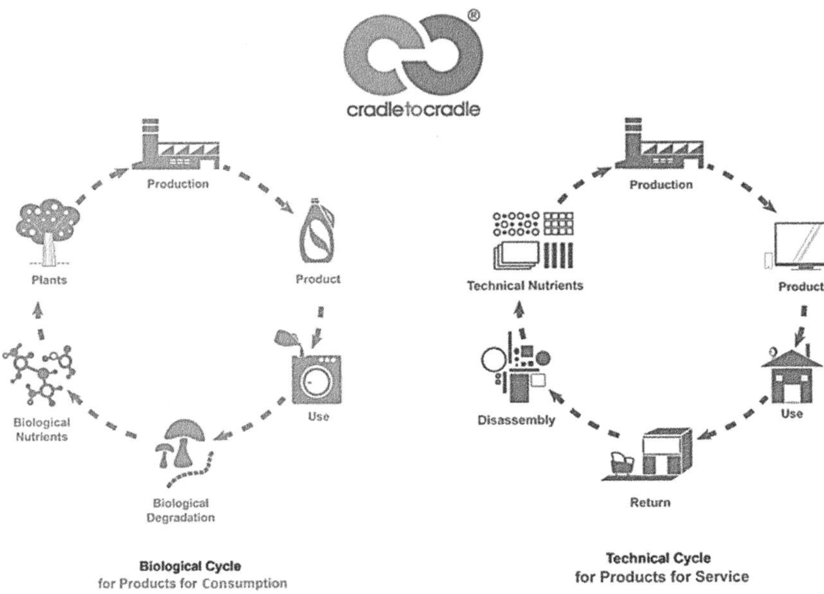

Figure 6.2. *Cradle to Cradle, between biological cycle and technical cycle (source: Upcyclea from MBDC[3])*

The technical cycle consists of intentionally controlled material flows. The idea is to allow mass industrial products to circulate in a closed circuit, while maintaining a constant level of quality. The fact that the system is closed is a prerequisite for creating optimal value. In a closed cycle, the use of toxic substances (if unavoidable, and until substitutes are found) can be controlled. The ease of disassembly and the care taken in choosing the materials for a product are fundamental aspects of the design. The products and materials in this cycle are called "use products". An object produced by assembling elements of defined quality can, for example, be reused to recreate the same material (closed loop) or participate in the manufacture of other products (open loop).

The technical loop also makes it possible to save functionality. This is, for example, the case of photocopiers where consumers do not buy the machine but use it and pay for the number of copies made. In this approach, the equipment remains

3 https://www.upcyclea.com/cradle-to-cradle.

the property of the manufacturer and is returned to the manufacturer after a defined period of use. One of the advantages of this system is that the manufacturer has an interest in using materials of a higher level and quality, in promoting repairability and dismantling in order to easily recover and reuse spare parts for which they have considered, during eco-design, the following uses, within economically resilient ecosystems.

6.1.3. *Counter false good ideas*

6.1.3.1. *Do not confuse recycling with the circular economy*

The concept of the circular economy is often associated with recycling. However, as our production models currently work, it is generally considered that we must "better manage our waste" by recycling it. However, when the products have not been designed to maintain the value of the material, recyclers cannot perform miracles.

Recycling is the "best" way to treat waste, whereas Cradle to Cradle eliminates the notion of waste and only deals with resources and their different uses. Recycling is an important first step, but it faces different limitations because there is a loss of value as the recycling process progresses.

6.1.3.2. *The false good idea of energy recovery*

Beyond recycling, we often talk about material recovery when we talk about incineration. It is important to give some details here. Energy recovery means burning materials which, although today there are no recycling channels for them, nevertheless means losing precious and rare resources, burning wet materials such as bio-waste and therefore using a lot of energy to create energy. It is time to change and to design differently.

6.2. Creating resilient models

6.2.1. *High-performance business models*

Constituting systems capable of preserving the value and quality of the resources used is the necessary condition to make a circular business model viable. By designing "material banks", the company can develop without damaging the environment or its profitability since from now on "secondary raw materials" have value, can be resold and are not transformed into waste.

Digital technology has its place and hence UPCYCLEA has designed an intelligent material library to manage material flows and avoid unnecessary storage by detecting – upstream – the need for used material.

6.2.2. *The material passport, a key factor in traceability*

With recycling, waste is recovered if its composition is unknown and most of its ingredients have not been designed for the next use and can therefore be harmful to health and the environment. Hence, creating a "circular passport" for each product provides reliable information that is useful in determining the potential uses of materials. It describes for a product:

– its composition, in terms of its components and the materials constituting these components;

– its period of use, i.e. the period beyond which the product will have deteriorated to the point of becoming waste and losing any residual value;

– its next possible lives, whether it is reusing, repurposing, upcycling or recycling, if it is a next technical cycle, or composting, if it is a biological cycle;

– other useful information, such as its carbon footprint, its recycled content, or maintenance or dismantling instructions to facilitate its disassembly at the end of use.

6.2.3. *Promoting the abundance of renewable energies*

To be consistent, a Cradle to Cradle product and, more generally, a company whose strategy is based on the principles of Cradle to Cradle cannot rely on fossil fuels since their negative impact on climate change is clearly proven. Here again, let us base our actions on the way nature works: solar energy is available, infinite and does not emit greenhouse gases. Other renewable energies can also be used: wind energy, biomass, etc. Without an energy source, also C2C, the products designed will only be optimizations of the existing system.

In nature, the richer and more diverse ecosystems are, the more resilient they are to change. The way we manufacture products must be modeled on this, with the same genius for diversity and variety.

6.2.4. *A recognized approach*

The Cradle to Cradle methodology has a certification process validated by an independent authority: C2CPII (Cradle to Cradle Product Innovation Institute). The first feedback on C2C experience dates back to 1993.

Certification allows the analysis of products designed according to certain criteria, including:

– non-toxicity of materials;

– belonging to a biological or technological cycle, respect for water;

– compliance with social rules; or

– the principle of corporate social and environmental responsibility.

Figure 6.3. *The different levels of Cradle to Cradle certification*[4]

This certification is divided into five levels: basic, bronze, silver, gold and platinum, which make it possible to promote the progress approach. All these levels of certification are valid for two years, except the basic level, which remains valid for only one year. In order for the product to be recertified at the end of this period, the quality of the product must be maintained. If the quality of the product has been improved, it can obtain a higher degree of certification. Today, more than 2,000 products are "Cradle to Cradle" certified. The list of C2C products is available at www.c2ccertified.org

6.2.5. *A demanding approach*

Companies that engage in Cradle to Cradle certification must integrate the advantages and disadvantages of being pioneers. First of all, carrying out extremely detailed analyses to know the fine composition of each ingredient in its products requires time, patience and investment. Moreover, committing to this approach requires consistency in terms of research and development. Finally, although this

4 https://mbdc.com/how-to-get-your-product-cradle-to-cradle-certified.

label is internationally recognized, it is not well known by buyers and consumers and therefore does not – for the time being – constitute a marketing and commercial argument for the general public. However, it is one of the few labels to integrate social, environmental and even health issues with high standards, with a view to achieving a positive impact.

6.3. Some examples of C2C certified products

6.3.1. *Carpets*

In a traditional method, all the components (textile and underlay) of a carpet are glued together, making the qualitative reuse of secondary raw materials almost impossible. Glue also has negative effects on air quality and the health of the inhabitants. In this context, a used carpet will, at best, be recycled as road underlay, at worst, incinerated or landfilled, but the components will never remain in a loop. Recycling therefore delays the destruction of the material but does not eliminate it. The loop is not closed.

To remedy this, major carpet manufacturers such as Desso/Tarkett or Interface have been involved in a Cradle to Cradle approach for years. By designing a non-toxic carpet with separable overlay and underlay, they can create positive loops and reintegrate the material into their production circuit. This requires the implementation of a new business model to recover used materials after a period of use determined between the producer and the customer, and in terms of business model, it avoids the purchase of new materials.

However, it should be noted that it is not because the product is compostable that it is composted, nor that because a product can return to a technical or biological cycle that this actually happens. This requires profound transformations at different levels.

Certainly, it is possible for local authorities to organize return and material-processing circuits. However, companies can also change their business models by integrating the principle of functionality economy, for example. A carpet could thus be made available for the duration of use and contractually recovered by its manufacturer. The eco-designed carpet can then be disassembled and the various materials returned to their technical or biological cycle, thus becoming a secondary raw material again.

6.3.2. *Buildings*

According to ADEME[5], buildings generate around 40 million tons of waste per year, the vast majority of which cannot be recycled. We also know how sensitive the

5 https://www.ademe.fr/dechets-batiment-0.

issues of access to materials for the building industry are becoming. By designing "C2C Inspired" buildings, this nonsense can be avoided. It is a question of designing places free of any toxic substances and eco-designed as material banks, i.e. able to be dismantled or disassembled. At the end of use, each object or material reintegrates its original circuit or a parallel circuit for a future rebirth without loss of quality or waste of the nutrients that were used to constitute it. This principle is called "C2C Inspired" because, in a building, it is currently impossible to use only C2C labeled products. However, it is the approach that is taken into account.

There are many buildings in France and internationally designed on this principle. One of the pioneers in France was the rehabilitation of the Lainière de Roubaix. Dominique Givois, Director of the SEM Ville Renouvelée[6], explains:

> From its extractable foundations to its roof made of unpaved vegetable membrane and its inflatable meeting room, the *Maison du projet de la Lainière* is part of a building designed in cycles. Each element is not only biosourced but can be dismantled and reused elsewhere and in a different way in the future.

This allows us to move from managing future rubble to managing preserved resources. In order to make this possible, it is necessary to create "circular building passports" and integrate them into a C2C inspired material library. The company UPCYCLEA has designed this material library, which allows us, thanks to artificial intelligence, to optimize the use of used materials according to market needs. This is a new business model that can be created. The economic value generated over the life of the building is thus reversed from management costs to management profitability.

6.3.3. *Fashion*

Fashion is an illustration of the excesses of our consumer society, of the waste of resources with what is now called "fast fashion". Moreover, according to ADEME[6], the textile industry is one of the most polluting in the world. While the European Union prohibits the use of substances such as lead, nickel or azo dyes, which are particularly toxic substances, this prohibition is not in force in the rest of the world. For several years, Greenpeace, in particular, through its DETOX[7] campaign, has denounced these practices and raised consumer awareness.

As consumer awareness grows, we see the need to act emerge in brand speeches and strategies. Some are committed to reducing the impact of textile production:

6 https://multimedia.ademe.fr/infographies/infographie-mode-qqf.

7 https://www.greenpeace.fr/defi-detox-monde-ne-se-defiler.

polluting less, consuming less energy and natural resources, recycling better, etc. This is an essential first step, but can polluting less be satisfactory if we always pollute? Should we not go further and reinvent other forms of production and consumption whose purpose will not be to be *less* polluting but to be *good* for health and the environment, i.e. to have a positive impact.

Engaging in a Cradle to Cradle circular economy approach to garment design will involve selecting non-toxic components, choosing suppliers that respect the human rights and health of their employees, promoting the use of renewable energy and ensuring that the quality of water generated by the factory production is better than it was when it came in.

To create "positive loops", it is necessary to plan the next life of the garment by organizing its recovery, recycling, even composting (for a cotton product) or transformation. For example, tights can be transformed to recreate tights or used to make carpets. It may also be possible to extend its life span by providing a system of rental and sharing.

Let us look at a few examples. As early as 2013, Puma was one of the first brands to commit to a "Bring Me Back" program with several products: lifestyle sneakers (biodegradable), Puma jacket (recyclable), T-shirts (biodegradable), backpack (recyclable). To close the loop, Puma proposed, to its customers, returning the products after use in order to recycle them and create new ones. C&A, the world's largest consumer of organic cotton, has worked with its suppliers for several months on non-toxic and fully recyclable products, even compostable ones, according to the Cradle to Cradle principle. The first T-shirt was released in 2017 at a price of less than €10. Since then, C&A regularly offers new C2C ranges. Wolford, a specialist in tights and lingerie, has created fine, biodegradable and non-toxic lingerie. Finally, the English designer Stella McCartney, already very committed to animal welfare, has also committed to detoxifying her creations with the Cradle to Cradle approach.

To better work together and accelerate the evolution of the textile industry, brands, designers and manufacturers of fibers, dyes, etc. have decided to join forces around a "Positive Fashion" movement.

6.3.4. *Cleaning products*

Beyond fashion, brands of green cleaning products such as Rainett[8] or GreenCare[9], which belong to the Werner & Mertz Group or Method, have the

8 https://www.rainett.fr/fr/notre_demarche_responsable/le_groupe/index.html.

9 https://wmprof.com/fr/fr/nachhaltigkeit_1/cradle_to_cradle_3/cradle_to_cradle.html.

ambition of having a positive impact. All toxic substances are gradually removed from the products, thus avoiding water pollution. Packaging is designed to be fully recyclable and made from recycled plastic, while still being affordable in price, and more and more products in their ranges are C2C certified.

6.4. Conclusion

The road is sometimes long because it is demanding. The brands that commit themselves do so on a product-by-product basis. All these actions may seem like drops of water, but for these pioneers, it is a trend of the future and their responsibility to contribute to it.

Everyone, at their own level, can play a role in "being part of the solution" rather than remaining part of the problem, whether they are brands or consumers.

As consumers, we can, of course, favor organic, repairable, reusable products and, where they exist, C2C labeled products. In France, a C2C Community has been created, whose objective is to popularize this approach and bring together brands, local authorities and citizens committed to the same objective, available at www.C2Ccommunity.org.

Cradle to Cradle contributes to building a world where we manage our resources rather than our waste, where we create and consume healthy, rather than less toxic, products. It is a technically very successful approach and certification, but today it is not very well known despite its pioneering nature. To become widespread, strong communication by all Cradle to Cradle actors seems necessary. It is time. There is still time.

6.5. References

Braungart M. and McDonough W. (2011). *Cradle to Cradle : Créer et recycler à l'infini*. Gallimard, Paris.

Braungart M. and McDonough W. (2016). *UPCYCLE, Au-delà de la durabilité, concevoir pour l'abondance*. Gallimard, Paris.

Jackson T. (2010). *Prospérité sans croissance, la transition vers une économie durable*. De Boeck, Brussels.

7

From the Circular Economy to Industrial and Territorial Ecology Approaches: What Modes of Governance to Ensure their Sustainability?

7.1. Introduction

Industrial and territorial ecology (ITE) is a concept strongly linked to the circular economy (CE) (Murray *et al.* 2017; Fernandes and Kadio 2017). It is now firmly established in the international, European and national landscapes in particular, mainly through strong government action. However, the initiatives taken and led by companies, associations and local authorities raise questions, particularly with regard to their sustainability. The question of the future of industrial and territorial ecology approaches within the framework of the CE thus becomes crucial. In this chapter, we focus on identifying the modes of governance of these approaches to ensure their development in the medium and long terms, especially when several of them coexist on the same territory. To this end, the first section presents the concepts as well as the levers and obstacles to the sustainability of ITE approaches. Then, the second section identifies the different forms of governance that exist through an exhaustive review of the approaches on the national territory. Finally, the third section, based on empirical research – the case of three ITE approaches in the territory of La Rochelle – allows us to question their future through their coordination and even their common governance.

Chapter written by Valérie FERNANDES.

7.2. Leverages and obstacles to the sustainability of ITE approaches: findings and analysis

7.2.1. *Circular economy and ITE approaches: definition and key elements*

The concepts of the circular economy and industrial and territorial ecology are intrinsically linked. They share the same fundamentals – reducing the environmental impact of industrial and economic activities – and their borders are difficult to identify (Murray *et al.* 2017; Fernandes and Kadio 2017). For Buclet (2015), "there is no doubt that industrial ecology and circular economy describe the same reality, refer to the same objectives and the same desire to increase the sustainability of industrialized societies." However, a more detailed analysis makes it possible to identify different levels of action, objectives and tools (Table 7.1).

Circular economy	Industrial and territorial ecology
A global approach at national or European level	A local approach at the scale of a territory
A primary legislative vocation	A primarily operational vocation
An economic vision of resource management	An economic and participatory vision of resource management
No clearly stated methodology	Industrial and territorial metabolism focused on quantitative aspects

Table 7.1. *Circular economy and industrial ecology, complementary approaches, adapted from Buclet (2015)*

This comparative approach makes it possible to affirm that ITE is today a field that is part of the circular economy[1], in that it represents a set of exchanges of flows (materials, water, energy) between different companies located nearby on the same territory. Its objective is to reduce the environmental impact of industrial activities (Erkman 2004; Ehrenfeld and Getler 1997; Varlet 2012; Beaurain *et al.* 2017). This approach is a major lever for the energy transition and is enshrined in the law of August 17, 2015 on the energy transition for green growth. The latter defines ITE as aiming, "on the basis of a quantification of resource flows, and in particular materials, energy and water, to optimize the flows of these resources used and produced at the scale of a relevant territory, within the framework of actions of

1 ITE is thus clearly mentioned – and repeatedly – in the circular economy roadmap published by the Ministry of Ecological and Solidarity Transition on January 31, 2019. Available at: https://www.ecologique-solidaire.gouv.fr/sites/default/files/Feuille-de-route-Economie-circulaire-50-mesures-pour-economie-100-circulaire.pdf.

cooperation, pooling and substitution of these resource flows, thus limiting the environmental impacts and improving the economic competitiveness and attractiveness of the territories."[2] It is therefore a question of moving from a linear type 1 ecosystem, where raw material resources are considered unlimited and the environmental impact of waste is not taken into account, to a type 2 ecosystem where waste is limited, or even a type 3 ecosystem where ultimate waste does not exist since it is reintegrated as a raw material. These three types of ecosystems (after Allenby 1992, cited by Gallaud and Laperche 2016) are represented in Figure 7.1.

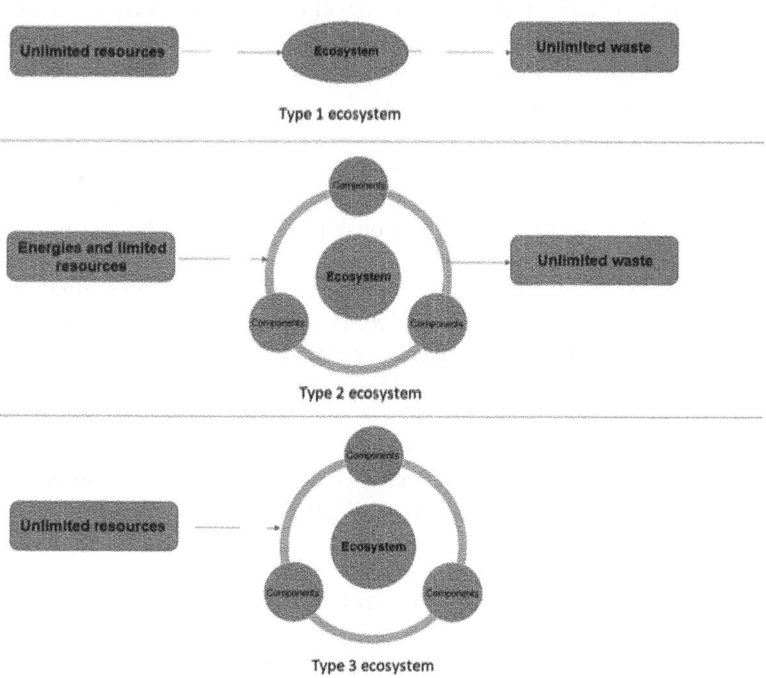

Figure 7.1. *Industrial ecology: different types of ecosystems (Gallaud and Laperche 2016, after Allenby 1992)*

The type 3 ecosystem is the ideal towards which ITE approaches tend to work, knowing that total looping is impossible due to thermodynamic laws (Gallaud and Laperche 2016).

Chertow (2007) identifies the main opportunities for resource exchange as the reuse of by-products as a substitute for raw materials or commercial products,

2 https://www.ecologique-solidaire.gouv.fr/lecologie-industrielle-et-territoriale.

the sharing of services or infrastructure and the pooling of services such as transportation or food. There are many reasons for an ITE approach: commercial reasons to reduce costs or increase revenues, long-term resource security through the provision of critical resources such as water, energy or certain raw materials. However, other benefits can also be generated, such as economic development, pollution reduction, water and land saving or greenhouse gas reduction. Finally, the interest of these approaches also lies in the revitalization of urban and rural sites, growth and/or job retention and, of course, more sustainable development.

ITE therefore makes it possible to develop links between local authorities, companies and other stakeholders in the territory in dealing with common problems. This local level makes it possible to implement concrete actions aimed at reducing negative externalities on the environment.

This territorialized network – like a competitiveness cluster even if the origin and context of their birth is different – goes through different stages of development, which often result in changes in the relationships between actors.

Today, approximately 100 ITE approaches have been identified in France according to a study by Duret *et al.* (2018). However, they have different degrees of activity: of the 84 approaches listed by the Orée association (2016), 10 are suspended, 59 are in progress and only 15 are considered sustainable.

This is because many challenges need to be addressed with regard to the sustainability of governance, operational sustainability and economic sustainability of these approaches (Duret 2005; Duret *et al.* 2018).

7.2.2. Success factors and obstacles to the sustainability of ITE approaches

7.2.2.1. Factors for success

The success and sustainability of industrial ecology approaches is today a crucial issue that can be analyzed from an operational, organizational or economic perspective (Duret *et al.* 2018). The various success factors are based on different elements such as the natural resources available, the facilities and infrastructures of the territory, the different economic sectors located in the territory, the quality of the relations between the different actors in terms of cooperation or the strong membership of the actors in the territory. The recognized legitimacy of the facilitator is also a fundamental attribute in the success of these approaches. Indeed, the latter, as an actor, has all the skills to successfully manage and sustain the process. He/she must be able to seek synergies; take action to promote the economic development of the territory; integrate entrepreneurial activities; train and transfer his/her skills; communicate, evaluate and promote the results; and organize convivial times between companies in order to create knowledge and mutual trust to consolidate the approach. The facilitator – who can conceive himself/herself as a third-party actor in the sense that he/she is the initiator, facilitator and guarantor of the trust and cooperation of the different organizations involved in the process (Geindre 2005) – must also reopen new strategies and new synergies. He/she must diversify his/her services to be more attractive to companies (pooling of resources, responsible purchasing, etc.). Some initiatives, such as that of AIRM (*Association d'Industriels de la Région de Meyzieu*), offer its members services on waste collection, security, management of green spaces, cleaning of industrial premises, purchase of supplies, among others. Others offer consulting and audit services in renovation, such as the CAVEM[3] in Fréjus.

The recruitment of new members in order to find new synergies is also an essential factor of sustainability. This is the case with the BIOTOP approach in La Rochelle, which carries out diagnoses for each new member and updates flow data annually in order to detect new needs and thus propose appropriate offers.

The integration of industrial ecology into training courses from secondary school onwards, the contribution of researchers and students on projects to provide scientific validation, the promotion of the results obtained and the integration of industrial ecology into urban, regional and spatial planning strategies are also real levers to enable the long-term development of ITE.

Ensuring the convergence of interests between stakeholders to effectively operationalize synergies is another factor for sustainability, as is the association of ITE with the solidarity economy. Indeed, these two approaches are complementary, in particular through the principle of cooperation, networking and information sharing, as well as the search for the creation of activities and jobs.

Finally, integrating an ITE approach into a territorial sustainable development project can make it possible to mobilize economic actors, and in particular companies, more effectively.

7.2.2.2. *Obstacles to the sustainability of industrial and territorial ecology approaches*

The most recurrent obstacle to the sustainability of industrial ecology initiatives is the responsibility of the companies that are operational players in these initiatives. Duret *et al.* (2018) point out that the decentralization of major groups' decision-making centers, followed by a high rate of renewal of managers, prevents

3 VAR-ESTEREL – Méditérranée urban community.

companies from becoming territorially based over time and can then compromise their involvement in decision-making on the approach. Moreover, the priority of companies is the short-term return on investment and if no concrete results are achieved in the first few years, their involvement in the process is compromised (Duret *et al.* 2007).

Although the plurality of actors is important for the governance of ITE approaches, it can also be a source of blockage to their sustainability due to diverging interests. As Ehlinger *et al.* (2007) point out, a network under different forces can become complex and difficult to govern because of the large number of participants. Political authorities and the public defend environmental interests while companies defend economic interests. It is then necessary to find the right measure to reconcile these different interests (Beaurain *et al.* 2017).

In a situation of shared governance, conflicts of responsibility can arise as well as difficulties in making joint decisions between companies and communities. This can lead to poor collaboration and can compromise the sustainability of the approaches (Duret *et al.* 2018). Politics, image, usage and value conflicts that have existed or exist between individuals or organizations can also violate the sustainability of the approaches (Gobert and Brullot 2015).

In addition, the increase in the costs of animation, the change of animator or even the change of participating organizations can also negatively influence the sustainability of the processes (Duret *et al.* 2018). The provision of human resources from local authorities may then be necessary (Duret *et al.* 2007).

Obtaining financial resources to ensure the sustainability of industrial and territorial ecology initiatives is a real challenge. Significant sums are injected at their start-up and then in their operationalization phase. However, finding the means to ensure their continuity is a real problem. Indeed, ITE procedures receive 70 to 90% public subsidies (from ADEME in particular) for calls for expressions of interest over a defined period. This raises the question of their sustainability beyond two or three years, knowing that the level of maturity is reached after a cycle of four to six years. It sometimes happens that during the process of building the approach, funding stops. Stakeholders close to the approaches must then find different sources to ensure development (Duret *et al.* 2018).

To this end, various means of raising private or public funds are used. Some approaches opt for public funds such as "project" financing (region, ADEME, BPI, State services, FEDER, for example) and operating grants (local authorities attached to the project: community of municipalities, urban community, city, department, region). This is the case of the Montauban Tarn-et-Garonne Chamber of

Commerce, which mobilizes co-financing[4] for carrying out diagnoses, developing shared employment, promoting bio-waste and leading the process. Others choose private funds such as contributions, membership, service and sponsorship.

7.3. The modes of governance of ITE approaches

7.3.1. *Key elements*

Like a territorialized network, ITE approaches require good governance to ensure their sustainability, especially when network actors are in competition with each other to retain resources that are not transferable from the territory (Mendez and Mercier 2006). The management, regulation and governance of the territorialized network are therefore essential to avoid the refusal of actors to participate in exchanges by compartmentalizing their resources, isolating themselves and even leaving the network, which can weaken it or even lead to its disappearance (Ehlinger *et al.* 2007). The formalization of governance is a central element for this network to develop its own internal dynamics and in relation to external stakeholders. Individual actors (company managers) as well as collective actors then decide to delegate to the governance body a certain decision-making power, while having a right of control over the latter.

The governance of an ITE approach therefore depends on the development of cooperation between public administrations and private companies, highlighting new forms of partnership between socio-economic and political-administrative actors. It implies a collective behavior that aims to take better account of ecological constraints, through shared objectives, the implementation of concrete actions and even the sharing of data.

Five types of territorial governance can be distinguished: private governance, collective private governance, public governance, mixed governance and associative governance (Colletis *et al.* 1999). Private governance is a self-organized system whose coordination and resource creation mechanisms are managed by the dominant private actors to appropriate the local productive space. This mode of governance has no formal organization. Collective private governance, on the other hand, is a formal private institution bringing together private operators such as chambers of commerce, professional unions and private business clubs that promote the coordination of their strategies. Public governance applies to local public institutions that manage resources through the production of collective goods and services that can be used by all actors without rivalry or exclusion. These public institutions may be the State, local authorities or public research centers. Mixed governance consists

4 Ademe-DIRECCTE-SIRTOMAD – Communities of the municipalities of the Deux Rives.

of the association of the different types of governance mentioned above but with a predominance of public or private governance. Finally, associative governance is illustrated by the fact that the members of the network feel like they are part of a community of solidarity because of geographical proximity. The steering function is then distributed among members who all have equal decision-making power with exchanges based on trust and solidarity. This mode of governance is successful thanks to an institutional framework that manages regulatory mechanisms by intervening in the implementation of collective rules.

An analysis of ITE approaches present in France shows that the vast majority of them – particularly the oldest – have chosen governance by third-party actors with associative status. Indeed, the 1901 law association seems to be the most flexible configuration in that it allows for better multi-partner governance.

Table 7.2 illustrates the variety of governance modes through seven examples: ECOPAL, SYNEO, AIRM, PIICTO, CEIA, FE2I and CLES. Four criteria are used in the comparative analysis: the start date, the project leader, the type of actions supported by the approach and finally the governance model adopted.

7.3.2. *The modes of governance of ITE approaches*

The governance of ITE approaches is very crucial to their success and to the coordination between actors. The latter are identified according to their power (ability to impose their will on others), their legitimacy (recognition of a political or social order as an authority) and their interest (economic, environmental or political advantage that the results of the process can bring). This governance therefore requires a plurality and variety of actors, important principles for balancing decisions.

Two modes of governance are used in the management of industrial and territorial ecology approaches: shared governance and traditional governance (Duret *et al.* 2018).

Shared governance is based on the principle of co-responsibility and self-organization of participants, and more generally on collective intelligence. This shared governance approach is emerging in France.

Conventional governance is characterized by the establishment of committees, boards and general assemblies where participants vote by a show of hands on decisions taken by one or a group of people. This traditional mode of governance can reduce collective participation for a better evolution of ITE approaches. It is therefore a more "directive" governance than shared governance.

	Start date	Project leader	Type of actions	Governance model
ECOPAL/Dunkirk	2001	ECOPAL association	Various pooling, flow study, substitution	Multi-actor via the general assembly and the steering committee
SYNEO/Valenciennes	2007	Pôle Synéo association	Pooling of energy supply	General Assembly and Board of Directors
AIRM/Meyzieu	2012	AIRM association	Pooling of services, sustainable development, energy, mobility, maintenance of green spaces, collection of industrial waste	Multi-actor via the general assembly and the steering committee
PIICTO/port area Fos-sur-MER	2013	PIICTO Association	Consolidation of the existing industrial fabric and improvement of the attractiveness of the port area	Multi-actor via the general assembly and the steering committee
CEIA/Aube	2003	CEIA	Organization of a network of exchanges, information and incentives for the implementation of innovative industrial and territorial ecology projects. Creation of a shared logistics platform for fresh products or the use of recycled materials for the construction of a road section	Multi-actor via the general assembly and the steering committee
FE2I/Florange	2013	VALO & SCOP Ecota Conseil	Increase the competitiveness of companies, produce professional integration through the creation of solidarity-based jobs and optimize the environmental impacts of industrial activity	Board of Directors Organization via a dedicated association
CLES/Strasbourg	2013	GUP (*Groupement des Usagers du port de Strasbourg*)	The CLES approach aims to reconcile the economic development of the port area of Strasbourg (the port being the largest economic area in Alsace), the environment and its urbanization objectives	Several steering bodies: Strategic Committee, Technical Committee, Works Council, Advisory Committee

Table 7.2. *Some examples of ITE with an associated mode of operation*

In parallel with these two modes of governance, two others can be mobilized depending on the situation: operational governance and political governance.

Operational governance occurs when conflicts arise between the companies involved in the process, which do not affect the continuation of relationships in the network. This type of governance ensures that established rules are respected, transactions and exchanges run smoothly and are based on the trust that stakeholders have in each other. For persistent conflicts, the problem goes up the hierarchy, which finds solutions through coordination tools, commissions, reviews and audits.

Political governance can be mobilized in the event of a major conflict of great importance. The use of external political powers that have sufficient legitimacy and power to resolve the conflict is then necessary.

In conclusion, there is no single and perfect model of governance that can be applied to all approaches, as each territory has its own characteristics.

7.4. Articulation of ITE approaches on the same territory: the example of the La Rochelle territory

This empirical research is based on a qualitative methodology based on semi-directive interviews and document review. It was carried out as part of a research contract financed by the La Rochelle Territoire 0 carbone consortium in 2040[5]. Two groups of actors were asked to participate: ITE activities[6] on the territory of La Rochelle, on the one hand, and institutional actors or those working within the scope of ITE, on the other hand[7].

A total of 10 interviews were conducted. They were recorded, transcribed and analyzed using the thematic analysis method.

7.4.1. *Presentation of the three approaches*

Three ITE approaches coexist on the territory of La Rochelle (extended to Rochefort, a city about 30 kilometers away) and have their own specific

---

5 This consortium is composed of the La Rochelle Urban Community, the Grand Port Maritime, the University of La Rochelle and Atlantech.

6 BIOTOP, MER, CIRCUL'R.

7 ADEME, Grand Port Maritime de La Rochelle, La Rochelle Urban Community, Pôle Eco-industries, Union Maritime, Région Nouvelle-Aquitaine.

characteristics. These are BIOTOP, MER and CIRCUL'R. They all benefit from traditional governance and have an associative structure.

BIOTOP is the most mature approach. Created in 2010, following group purchasing actions within the Périgny business club, the SPHERE association was created, BIOTOP being the brand[8]. In order to start its activity, BIOTOP benefited from the financing of the region and the eco-industry cluster, then of the CDA (*Communauté d'agglomération de La Rochelle*) and the ADEME. Today, BIOTOP has found its business model, self-financing about 80% of its activity, the only public subsidies received today being those from CDA. Two major activities structure the approach: training and waste management, in the sense of pooling flows and substitution synergies. For example, the waste from a roaster (jute bags) is now being reused as a substrate for the neighboring nursery grower. More than 60 flows have been identified by BIOTOP. There are 112 members, the majority of whom are located in the Périgny area.

Matière Energie Rochelaise (MER) started at the beginning of 2016, with the recruitment of a facilitator in May of the same year. This ITE approach is an emanation of Port Atlantique La Rochelle, which responded in 2015 to a call for expressions of interest from ADEME and the Poitou Charentes region. Located on the port square, it has 40 companies involved (or at least identified). Governance was provided by the Grand Port Maritime de La Rochelle (GPMLR) and steering by a committee composed of the financiers (Région Nouvelle Aquitaine, ADEME), CDA, the Maritime Union and EXCELIA Group until 2019. Since then, MER has become independent, with a legal structure of the association type under the 1901 law. Today, the companies involved in the process are required to pay a contribution. ADEME and the region finance 70% of the post in the first year, 60% in the second year and 50% in the third year. These seed funds therefore expired at the end of the third year, in May 2019. The specificity of MER is its port nature, as well as the fact that it "goes on all possible flows": energy, water, waste, services and even a methanisator project.

Although its application was not accepted by ADEME during the call for expressions of interest launched in 2015, CIRCUL'R was created in July 2016, in the form of an association under the 1901 law. However, ADEME allocates funds for carrying out the diagnoses. CIRCUL'R then calls on two AMOs (project management assistance): the Rochefort Chamber of Commerce via a dedicated project manager and BIOTOP. A total of 40 companies were identified, each of which contributed financially to the diagnosis. At the end of 2018, CIRCUL'R successfully responded to a new call for expressions of interest from ADEME and the New Aquitaine Region in order to finance a facilitator.

8 For ease of reading, we will continue to use the name BIOTOP.

7.4.2. Challenges, risks and opportunities

7.4.2.1. Specific ecosystems

The representation of the ecosystems of the three approaches highlights their degree of proximity to each other and to other stakeholders, particularly institutional stakeholders. Thus, BIOTOP and CIRCUL'R are very close, not only because of the relationships established but also because of the very structure of their business model, while MER is perceived as specific, because of its port positioning. With a few rare exceptions, the participation of companies in one of the three approaches is exclusive. Indeed, it should be noted that the typology of companies varies according to the approach: those of BIOTOP or CIRCUL'R are more VSEs and SMEs while MER has more large structures and the land belongs to the GPMLR.

While the main institutional partners are part of the ecosystems of each approach, they are not involved to the same degree.

ADEME is thus naturally more present in the approaches it currently finances, namely MER (year 2 of financing, presence in the steering committee) and CIRCUL'R (via the financing of part of the audits until 2018 and the seed fund at the end of 2019). It is less so for BIOTOP, since its financing was spread over the period 2010–2013.

The eco-industries cluster (pôle éco-industries, PEI)[9] has been involved since the start of BIOTOP. Thus, over the first three years of BIOTOP, the IAP provided the equivalent of 0.3 FTE (full-time equivalent) per year. While in 2010 it had prerogatives in terms of funding, this is no longer the case. However, it continues to have ongoing relationships with BIOTOP. The same applies to MER, where the IAP followed the structuring of the process and provided technical assistance. Relations with CIRCUL'R are more relaxed due to the lack of regular meetings.

The *Communauté d'agglomération de La Rochelle* is historically very close to BIOTOP, the first ITE approach on the territory. It has been supporting it financially since 2013 (approximately 30% of BIOTOP's total budget). It is also present at MER, in that it participates in the steering committee and more broadly through the financial support it provides to the port area. However, it has no link with CIRCUL'R.

---

9 The eco-industry cluster was created in 2005 at the initiative of the Poitou-Charentes region to support, accompany and promote the development of the eco-industrial sector. Its mission is also to create innovative companies that reflect the values of the circular economy and energy transition.

7.4.2.2. *An institutional environment that raises questions*

It should be noted that the various institutional actors mentioned above are all questioning the future of ITE in general and of certain funding methods in particular. More specifically, they question the articulation they consider desirable between the different ITE approaches in the same territory, in order to overcome the anticipated risks. These questions all stem from a repositioning of the actors on their fields of action: who can finance what, who will finance what, according to what expectations?

The CDA, ADEME, PEI and the New Aquitaine region are all part of this reflective process. While the angles of analysis may differ, the same opinion is generally shared: the roles of the various actors must be clarified, and the three ITE approaches must be articulated. The objective is to ensure the sustainability of the initiatives and to establish La Rochelle (extended to Rochefort) as a territory of excellence in this field.

Thus, both ADEME and the region would like to develop a real "regional ITE center". As one of the interviewees put it: "I see these three approaches as an entire entity but there are interactions to imagine between the three." Because the territory of La Rochelle is truly atypical, with three approaches: "few territories at the national level that can pride themselves on having such successful and ambitious approaches and with a vocation for sustainability." It seems that in the short term there is a need to clarify each other's missions in order to identify bridges between the approaches. In the background, we can also identify within the various bodies interrogating the willingness to encourage these approaches to collaborate with each other in order to reduce the risk that could weigh on their sustainability in the event of a decrease in public subsidies. The legislative context has thus led CDA to sign an agreement with the New Aquitaine region to continue to finance BIOTOP, in accordance with the NOTRe law, which reinforces the region's role in economic development and restricts the economic intervention of municipalities or communities of municipalities. There is therefore no guarantee that this agreement will be renewed in the same way in the future. In addition, the CDA, which is seeing the evolution of ITE approaches and the development of initiatives, is ultimately questioning the relevance of funding only one structure with regard to the other two. This is all the more so as the Economic Development Department is well aware of the challenges of ITE development in terms of innovation and job creation. This raises the question of the political line and strategy that CDA will adopt in this area in the coming years. Perceived competition between local approaches is therefore to be avoided at all costs.

Another risk is that of a divergent perception of the ITE, leading funding actors to contradictory actions, which could jeopardize the long-term sustainability of ITE approaches. More precisely, two visions are opposed. The first is to adopt a minimalist and pragmatic viewpoint, which focuses mainly on dyadic actions between companies. In this context, the sustainability of relations between the two companies is far from being guaranteed because there is no common and territorial project bringing together several economic actors. The second vision is based on the creation of an eco-network of local companies, which then share a project for the future on the same territory. These two visions are contradictory in terms of both methodological tools and financing methods. There is a risk that two funders will each adopt one of the two visions.

In addition, the robustness of business models varies according to the approaches studied. The degree of maturity is the major explanatory factor. But so is the development of financial resources. MER has just changed, with external financial resources having to be found to compensate for the end of the ADEME seed fund: these resources must come from both companies and local authorities (including ports). More broadly, while it is indisputable that ITE approaches must develop private resources, it is very important to maintain public funding, or else they risk endangering their sustainability. Indeed, maintaining private financing over a medium term is extremely delicate if the latter are linked to high ROI actions, but ultimately it is not very ambitious in terms of environmental protection.

Finally, a final risk is identified: that of a distance between the BIOTOP and CIRCUL'R approaches to the MER approach, due to a very strong organizational and cognitive proximity between the first two approaches, hindering a substantial rapprochement with the third.

7.4.2.2.1. Strong challenges for the territory

The challenges and projects currently underway in the La Rochelle region concern the achievement of the 0 carbon objective in 2040, via the 0 carbon territory approach in 2040. It is also a question of weighing up against the regional capital in order to obtain funding, especially since ambitious initiatives are being taken in Bordeaux, as well as – above all – that the regions see their role in leading, structuring collective actions and funding strengthened. In other words, how can communities of municipalities, such as CDA in particular, obtain the support of the region to develop ITE actions on their territory? The need to structure approaches and initiatives is therefore becoming increasingly urgent. "It is the ability of the agglomeration to understand that these approaches are a territorial project that is part of economic development." CDA must be able to define a strategy in this area. And to do this, a number of questions are asked: what evaluation of the approaches? On the basis of which indicators (e.g. recycled waste, jobs created)? Should ITE be extended (and if so, how?) to the other business areas of the La Rochelle agglomeration (12 in total)? How can we encourage the creation of new sectors? Is there a need for a global structure to coordinate the three existing and possible future approaches?

7.4.2.2.2. A convergence of governance modes

In view of the challenges raised above, the minimum development of a common approach between BIOTOP, MER and CIRCUL'R is not debated by the various stakeholders interviewed. However, the question arises as to the content of this approach, its form, its mode of governance and organization as well as the objectives pursued. The sustainability of this approach is based on the creation of a network, governance and a territorial project.

A merger of the three approaches will enable ITE in La Rochelle to be recognized at the regional and national levels, thus encouraging public subsidies, such as labels, to be obtained. Through synergies and pooling actions to be identified, it will be necessary to strengthen innovation (through new sectors for example) and job creation. Finally, such a rapprochement can only serve and be consistent with La Rochelle Territoire 0 carbone 2040.

Two approaches are being studied: the networking of the three ITE approaches and the creation of a cluster, to eventually allow a spin-off on the territory.

Networking is now in place, with BIOTOP and MER initiatives having increased their exchanges. Through a better knowledge of existing synergies, identified and processed flows, it is a question of identifying the skills and needs of each one in order to implement possible cooperation (feed flows, development of new synergies in other territories, etc.).

The creation of a cluster is not a topical issue but would make it possible to structure this initiative in order to guarantee its sustainability and development. It would be a question of creating an "innovative" cluster in ITE, according to models already existing in other sectors[10]. The objective is to make it possible, through an official structure (but with collegial governance), not only to identify an ITE entity in La Rochelle but also to generate public funds in order to allow a real development of innovative sectors in this field.

To our knowledge, only one cluster in France is positioned in this field. This is the "Circular Economy and Recyclable Materials Recovery" Cluster (Lot et Garonne), which brings together 20 organizations (companies and public bodies).

10 Examples of innovative clusters: Silicon Valley, Sofia-Antipolis, Futuroscope.

7.5. Conclusion

ITE approaches, rooted in the circular economy, are nowadays essential in the face of the environmental and climate challenges we face. They are unique in that they make it possible to achieve objectives quickly through the networking of companies and organizations in the same territory, requiring both geographical and organizational proximities. As exciting objects of study and research, ITE approaches are not at the very least fragile, as they are based on an unstable dynamic between public and necessary private funding and on sometimes contradictory objectives. Their future is crucial to achieving the CO_2 emission targets that France and Europe agreed in the Paris Agreements in 2015. The mode of governance is therefore an essential question, which arises even more saliently when it appears necessary to coordinate different ITE approaches in the same territory, as a condition for their development. While these concerns in this area may seem quite new, it is likely that the short-term future will see a proliferation of initiatives similar to those of La Rochelle.

7.6. References

Beaurain, C., Maillefert, M. and Varlet, D.L. (eds) (2017). La proximité au cœur des synergies éco-industrielles dunkerquoises. *Flux*, 3(109–110), pp. 23–35.

Buclet, N. (2015). *Écologie industrielle et économie circulaire : définitions et principes. Économie circulaire et écosystèmes portuaires*. Éditions EMS, Caen, pp. 27–41.

Chertow, M.R. (2007). "Uncovering" industrial symbiosis. *Journal of Industrial Ecology*, 11(1), pp. 11–30.

Colletis, G., Gilly, J.P., Leroux, I., Perrat, J., Rychen, F., Zimmermann, J.B. and Pecqueur, B. (eds) (1999). Construction territoriale et dynamiques productives. *Science de la société*, 48, pp. 25–47.

Duret, B. V.C., Mat, N., Bonard, A., Dastrevigne, E. and Lafragette, A. (eds) (2007). Ecologie territoriale : une aide à la définition d'une politique énergétique. Comprendre l'économie physique du territoire. *Les annales de la recherche urbaine*, 103, pp. 73–78.

Duret, B.V.C, Des Gayets, M. and Dain. A. (eds) (2018). Pérennité des démarches d'écologie industrielle et territoriale en France : constats, enjeux, et recommandations. Report, Auxilia, Pôle Eco-industrie, Médiane.

Ehlinger, S., Perret, V. and Chabaud, D. (eds) (2007). Quelle gouvernance pour les réseaux territorialisés d'organisations ? *Revue française de gestion*, 170(1), pp. 155–171.

Ehrenfeld, J. and Gertler, N. (1997). Industrial ecology in practice: The evolution of interdependence at Kalundborg. *Journal of Industrial Ecology*, 1(1), pp. 67–79.

Erkman, S. (2001). L'écologie industrielle, une stratégie de développement. *Le débat*, 1(113), pp. 106–121.

Fernandes, V. and Kadio, C. (2017). Impact de l'économie circulaire dans le management de la supply chain : une étude exploratoire. *Logistique & Management*, 25(4), pp. 15–25.

Gallaud, D. and Laperche, B. (2016). *Circular Economy, Industrial Ecology and Short Supply Chain*. ISTE Ltd, London, John Wiley & Sons, Hoboken.

Geindre, S. (2005). Le rôle de l'acteur tiers dans la construction d'un réseau stratégique. *Revue française de gestion*, 154(1), pp. 75–91.

Gobert, J. and Brullot, S. (2015). L'écologie industrielle et territoriale : une nouvelle logique d'organisation des acteurs et de mobilisation des territoires. *Lucrările Seminarului Geografic Dimitrie Cantemir*, 38, pp. 39–54.

Mendez, A. and Mercier, D. (2006). Compétences clés de territoires – le rôle des relations interorganisationnelles. *Revue française de gestion*, 164, pp. 253–275.

Murray, A., Skene, K. and Haynes, K. (eds) (2017). The circular economy: An interdisciplinary exploration of the concept and application in a global context. *Journal of Business Ethics*, 140, pp. 369–380.

Orée (2016). Le recueil des démarches d'écologie industrielle et territoriale. Report, Orée.

Valet, D. (2012). Enjeux, potentialités et contraintes de l'écologie industrielle : le cas de Dunkerque. PhD thesis, University of the Littoral Opal Coast, Dunkirk.

8

Sharing Economy, a Driving Force of the Circular Economy?

8.1. Introduction

After stating that we are what we own, Russell Belk (1988) now argues that we are what we share (2014a): in more than 20 years, Web 2.0 has radically changed our consumption practices. Bartering, exchange, sharing, co-production, co-creation and shared goods are realities once reserved for interactions between relatives, and are now widespread between strangers connected via platforms. Possession is gradually replaced by use in a new paradigm of the economy of functionality, which can be summarized as "the substitution of the sale of the use of the property for the sale of the property itself" (Burg and Buclet, 2005, p. 29). Why own a drill when studies show that it is used less than one hour a year when it can be lent, rented or shared?

While some uses of this new so-called sharing economy make it possible to optimize the use of goods, they can go hand in hand with profits and do not necessarily imply that they are free of charge or environmentally friendly. This sharing economy is linked to the circular economy (as defined in Chapters 1 and 2), which aims, from the design stage, to make better use of natural resources and preserve the environment. The sharing economy and the circular economy sometimes overlap marginally. Currently, mainly due to consumers convinced by the ecological urgency, the circular economy has not yet spread massively. It is facing Moore's chasm (1991), which it is struggling to cross. The purpose of this chapter is to provide a clear picture of new consumer practices and the motivations and barriers that motivate consumers who have converted to the shared economy to explore the potential leverage that these practices provide for the circular economy. The managerial implications for industrial players and public authorities will follow.

Chapter written by Catherine LEJEALLE.

8.2. Web 2.0 disrupts consumption practices

8.2.1. *Arguments over definitions*

Web 2.0 provides features to connect Internet users, to create or share content (e.g. Wikipedia) or exchange goods (e.g. LeBonCoin) or information (e.g. TripAdvisor) (Carroll and Romano, 2011). In comparison, Web 1.0 only allowed one-way connections such as from an information site to subscribers, without the latter being able to interact with each other or even respond to the information site.

Many companies were created to exploit the unprecedented potential of Web 2.0: AirBnB, BlaBlaCar, Youtube, Flickr, etc. Some make profits, others do not. The term "sharing economy" was very quickly coined, but according to the authors, the definitions vary. Acquier *et al.* (2016) compare about 10 definitions to conclude that it is an umbrella concept that covers a wide range of practices and business models. Some authors include gifts and bartering in the sharing economy, while others exclude them. In addition to the lack of a consensus on the definition, the concept is widely contested because *in the end*, sharing is non-existent or reduced. Belk (2014b) speaks of semantic confusion.

While for most authors, such as the Directorate of Competition, Consumer Affairs and Fraud Control[1], sharing economy and collaborative economy are synonymous, Belk (2014a) sees a difference. According to him, the collaborative economy consists, for some people, of coordinating online the acquisition and distribution of a resource for an amount or against compensation. If there is no compensation, he describes it as a sharing economy, but not a collaborative economy. It includes non-financial compensation, so that it considers gifts, bartering and exchanges as part of the collaborative economy. However, he specifically excludes CouchSurfing, the site where Internet users invite others to stay on their couch without financial compensation or any commitment to reciprocity. Concerning the sharing economy, Belk (2014a) distinguishes true sharing from four forms of pseudo-sharing. The differences are related to financial motivation (profit or not), the expectation of reciprocity and the existence of a community. He thus theorized different forms of sharing.

Belk (2007, 2010) also defines the sharing economy in comparison to the traditional economy based on monetary exchange.

1 https://www.economie.gouv.fr/dgccrf/Publications/Vie-pratique/Fiches-pratiques/economie-collaborative.

The main characteristics that oppose them are:

– reciprocity: if not systematic, it can exist in the sharing economy. It is totally absent in traditional economics, which is based on a financial transaction with payment;

– money: money, necessary for traditional purchasing, can be replaced by free, donated, exchanged services or bartering in sharing;

– social bond: sharing can create a social bond, but not necessarily, if it is done through a platform that respects anonymity;

– shared ownership: responsibility can be shared when users share a good;

– dependency: sharing induces dependency between users to set a schedule for the availability of the shared good, which does not exist during a pure purchase, where the owner is unique.

He thus describes a continuum of practices that range from pure sharing (a mother's care for her child) to commercial exchange (buying bread at the bakery). Figure 8.1, adapted from Belk, presents this continuum.

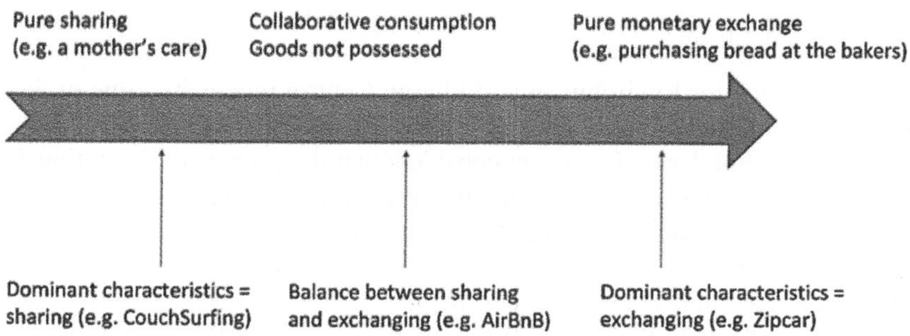

Figure 8.1. *Continuum between pure sharing and pure exchange*

Platforms where sharing is based on shared ownership, free access and a social link, such as CouchSurfing, are close to pure sharing. Platforms like AirBnB that have a balance between sharing and exchange characteristics are in the middle. Platforms like Zipcar with a dominant exchange characteristic are close to it. Habibi *et al.* (2017) define a scale to assign a score to each sharing solution in order to situate it on this continuum between pure sharing and commercial exchange.

8.2.2. *Mapping the sharing economy*

In order to better understand what it covers, Acquier *et al.* (2017) propose a model that organizes the sharing economy into three pillars, with intersections. These three pillars are: the access economy, the platform economy and the community-based economy. Each one offers different promises and leads to tensions and paradoxes.

The access economy consists of sharing goods and resources that would otherwise be underutilized in order to optimize their use. In contrast to the sale of property that involves a transfer of ownership, this involves a company or individual charging for the use of property and services, of which they retain ownership. Thus, Michelin Tire Manufacturer no longer sells tires, but miles driven, to trucks. Clarlight's company is offering to buy light, not fixtures. And, since 2006 and 2007, respectively, Spotify and Deezer have been marketing music subscriptions and not music media. This is also the case for carpooling with Autolib in France or Zipcar in the United States, Canada and Turkey, but it also applies to housing and swimming pools, luxury clothing and accessories, gardening and DIY tools or household equipment (drills, chainsaws, washing machines, etc.). It leads to a post-possession society where use takes precedence over possession and brings a benefit to the environment.

If the Agency for Environment and Energy Control retains it as one of the seven points defining the circular economy (see Chapter 2) and if the Ministry of Ecological and Solidarity Transition considers that the economy of functionality is a first step towards the circular economy[2], the researcher Tukker (2004) expresses reservations by indicating that there are tensions between economic and environmental incentives. Not being owners of the good, in the absence of social control of a community and a lack of reciprocity, consumers do not take as much care of the goods as if they were their own, or even make often inappropriate use of them. Establishing effective control mechanisms is costly and often prohibitive. In the end, the life span of the assets is shorter and the environmental benefit questionable, or at least complex to assess. Indeed, a shared electric scooter offered by operators such as Lime, Dott or Bird has a life span of about three months, compared to the personal scooter that its owner takes care of, which can last for years. However, the shared scooter may have been used for many more hours than the scooter used by its sole owner. If this argues in favor of sharing, we must also take into account the trucks that move the scooters from accumulation points to where they are missing, and realize that the balance may not be so favorable to sharing. Finally, the quality of scooters is not comparable: one is manufactured in large series in Chinese factories to gain market share by being present everywhere,

2 https://www.ecologique-solidaire.gouv.fr/leconomie-fonctionnalite.

the other is of superior quality. This example clearly shows that the overall impact of use is more complex to understand than it seems.

Moreover, the Jevons paradox (1865) concerning the environment applies perfectly to this case. In the 19th Century, Jevons noted that technological advances and efficiency gains in steam engines did not lead to a reduction in coal consumption but, on the contrary, to an increase in consumption, due to an increase in demand. At a lower cost, coal becomes more accessible. Also known as the rebound effect, the paradox calls into question the environmental benefits: by reaching new consumers who until then could not consume for economic reasons, there is an increase in overall consumption.

The second pillar, known as the platform economy, is based on the creation of a platform that allows Internet users who did not know each other before but whose expectations meet. The platform opens up a new meeting of supply and demand. These platforms disrupt the economy by competing with existing players. The case of AirBnB, which puts individuals in contact with each other to rent housing, illustrates this. Critics accuse them of making profits without really sharing anything or creating a community, while exploiting their employees (Carboni, 2016).

The third pillar is the community-based economy. It refers to non-contractual, non-hierarchical, non-monetized initiatives of interactions within the framework of a project, a task to be performed or relational exchanges. The objective is not to optimize the use of a property but to create social cohesion, promote values and accomplish a mission. Missions often have a political or ideological significance. Linux and Wikipedia illustrate this pillar. Despite an egalitarian vision and a professed openness, critics believe that they are finally recreating an unequal hierarchical organization and producing social closure between people who share the same convictions (Bradley and Pargman, 2017). Wikipedia responds that there is a need for an entity that validates the knowledge spontaneously written by Internet users. This is a paradox of openness and equality because the facts undermine official discourses.

8.3. The circular economy in Moore's chasm

8.3.1. *The pillars of the circular economy and the role of consumption*

The circular economy is a new model that combines economy and respect for the environment in a global societal approach. Better consumption will contribute to economic, environmental and social performance (Bonet Fernadez *et al.*, 2014). This process of networking the actors of a territory will transform them and will also modify employment (Maillefert and Robert, 2014). However, the link between the circular economy and consumption practices is not addressed in the definitions. On the contrary, it seems important to approach the circular economy also from the point of view of consumer behavior. They have a role to play in the dissemination of virtuous practices.

8.3.2. *Adoption of the circular economy: consumer behavior*

Even if the circular economy is based on industrial ecology, which, from its conception and intention, regenerates the goods consumed, consumer behavior is fundamental. The adoption or rejection of the circular economy passes through them, as a relay or incentive for industrialists and public authorities. Considering the impact of consumers on brands, whether it is boycotting brands or changing the composition of products, it is impossible not to be interested in their point of view.

Circular economic practices are measured by large state agencies. Thus, the analysis of the characteristics of household consumption allows the *Centre de recherche pour l'étude et l'observation des conditions de vie* (Research Center for the Study and Observation of Living Conditions) (Siounandan *et al.* 2013) to distinguish "constrained frugality" from "voluntary simplicity". While the former is the result of economic constraints, the latter is chosen, claimed and often militant. Siounandan *et al.* thus draw up a typology of households in five categories, according to their behavioral profiles and consumption strategies. There are:

– the basics (18% of the population), that are not interested in consumption in general;

– the wealthy (20% of the population), indifferent to possible consumption strategies;

– the thrifty consumers (22% of the population), in search of a good deal;

– the strategists (21% of the population), clever at finding good deals;

– the constrained (14% of the population), economically vulnerable;

– the engaged consumers (14% of the population), who use bartering, opportunity, borrowing and generally choose frugal lifestyles.

Concerning the engaged consumers, households with two children are over-represented. Siounandan *et al.* note that they have not adopted a consumer behavior in response to the economic situation but in search of meaning. This survey shows that only 14% of the population is reported to be deliberately adopting behavior favorable to the circular economy. In addition, the household business survey[3] shows that, in 2014, one in two people declare that they pay attention to the

3 Figures and statistics no. 505, INSEE – SoeS, March 2014.

origin of the products or the quantity of waste produced, i.e. a significant increase in vigilance, but only 13% make it a systematic choice criterion. Parisians, rural people and the elderly are more concerned about environmental issues, by sorting and reducing waste.

8.3.3. *Moore's chasm to cross*

In terms of adoption of innovations, Rogers (2003, first edition published in 1962) divides the population into five segments, according to their propensity to adopt an innovation. Innovators (2.5% of the population) are the first to adopt, followed by early adopters (13.5%), the first majority (34%), the late majority (34%) and then the laggards (16%). Over time, following an S-shaped curve, the different segments adopt innovation. Concerning hi-tech products and disruptive uses, Moore (1991) observes that once the first two segments have been converted (16% of the population), innovation collapses. There is a gap in sales or the number of downloads. While the first two segments are fond of novelty and easy to attract, the same cannot be said for the following segments, which, cautiously, require proof of the additional benefit provided. In addition, this segment does not communicate with the previous ones. Crossing the chasm and ensuring the diffusion of innovation requires identifying niches for which innovation brings a perceived relative advantage and then communicating specifically towards each niche. Percolation between segments will follow.

The above figures on the uses of the circular economy show that its spread is before the chasm, with 14% of the population convinced. The relative advantage for activists is the preservation of natural resources, but this alone is not enough to convince other consumers. We have to find more. Regarding obstacles, theories on innovation show that the complexity of use has a discouraging effect. This is corroborated by qualitative interviews conducted between June and September 2019 with a sample of convenience: adopting circular economic practices "makes life more complicated". Some sites like WAG[4] clearly state this: "I don't want it to be difficult! I don't know where to start!" In addition to this lack of knowledge of real uses, there is also a lack of knowledge of applications: alongside some highly mediatized applications such as Yuka, those such as TooGoodToGo, avoiding waste, must be better known. The cost of a press relations strategy is not within the reach of all start-ups.

4 https://www.wwf.fr/agir-au-quotidien/we-act-for-good.

8.4. The uses of the sharing economy

8.4.1. *Motivations*

The authors cited above show that a possible contribution of the uses of the sharing economy can be the ecological impact. Car-sharing (BlaBlaCar) optimizes vehicle filling, while carpooling (CarTravel) optimizes vehicle use between drivers. However, neither carpooling nor car-sharing is considered to be part of the circular economy. On the other hand, they are practices of the sharing economy. The other two motivations for using the sharing economy are financial (a more economical solution) and social (creation of a link by belonging to a community).

A study of the motivations of the 900,000 members of Zipcar, a car-sharing company with 10,000 vehicles, shows that, contrary to the ideals and values of the platform, members have no desire to belong to a community and no desire to share activities. Their motivation is purely financial (Bardhi and Eckhardt, 2012). Despite Zipcar's efforts to promote a new, more ecological and social lifestyle, users have other motivations. This study suggests that the motivations are irrelevant, if, *in the end*, the practices lead to a positive impact on the environment.

Interviews with BlaBlaCar users in France confirm these results. An occasional driver on BlaBlaCar, Jean-François (telecom executive, 48 years old, 2 adult children), explains: "It was my children who enrolled me because I had to drive back alone from Lyon to Paris. An hour later, I had two passengers. I do it occasionally if I'm alone, between Paris and La Grande-Motte for example this summer. I loved meeting people who were very different from those around me. It also prevents me from falling asleep at the wheel with company and it also reimburses me for a portion of the gas." Chantal (nurse, 50 years old, 3 adult children) is sometimes a passenger: "I used to go to see my parents in Metz and the train from Aix les Bains is not practical. My boys signed me up and I thought it was very nice." The younger ones are more interested in the price. Joaquim (student, 22 years old) says it all: "It's unbeatable in terms of price, offers last-minute flexibility that the train doesn't have." Axel (student, 21 years old) concludes: "I travel more because it is much cheaper than a train or a plane. I've been able to go to music festivals because of it." This last verbatim quote illustrates the Jevons paradox mentioned above. BlaBlaCar encourages this young person to travel more, whereas the real ecological attitude would be to reduce travel. However, there is some argument that the driver would have made this trip by car anyway; it adds nothing. However, this leads to the emptying of trains, which then run when they are less full. Again, the overall impact is complex to assess.

The interview with the founders of the sharing economy platforms confirms the three possible motivations (financial, social and ecological). The founder of the platform for the exchange of services and goods between neighbors, AlloVoisins, Edouard Dumortier, says that his motivation when creating it was to create social links and optimize the planet's resources. The white paper[5] he published sets out this vision, but in practice, three motivations coexist:

> "Is it really necessary to own or buy new products to consume? After all, this drill you need so badly (to make only one small hole a year), does it really have to be yours alone? What about the Rraclette machine? In short, for all our daily needs, we prefer the occasional use to ownership. It is cheaper, less restrictive and more responsible. In a word: it makes sense! Neighborhood support is the future! But in reality, some people want to enhance their hedge trimmer, which is never used, and earn some money. Others want to meet neighbors to share a meal or an outing. Still others are clearly motivated by the environment. It doesn't matter. The environmental impact almost always results from this, even if it is not explicitly intended by the neighbors."

Braineet is a company with an ideation platform, open to all, to meet the challenges posed by companies. A challenge for the BHV is to think of a friendly place where customers can meet and exchange ideas. Jean-Philippe Lorinquer, its General Manager, describes the motivations of participants who submit ideas online:

> "The Braineet community is active through new and ongoing challenges but the motivation to earn is limited. For example, one year of shaving products. Participants are committed to sharing their innovative ideas, reacting to those of others and feeling useful."

The Yoyo start-up is an example of a strategy to accelerate participatory waste collection. A total of 25,000 families have already registered to collect plastic bottles by filling bags that they give away for a reward (cinema ticket, free transport, etc.) to volunteer leaders (building managers). Its founder Eric Brac de la Perrière explains:

> "Collaborative collection is effective – the rate is 90% versus 20% usually – because it creates social cohesion."

Reward and social connection are two motivations that are put forward rather than trying to convince with an activist discourse on ecology.

5 https://www.allovoisins.com/page/consommation-collaborative.

8.4.2. *Percolation of uses*

Because it brings direct benefits to consumers, the shared economy is booming (Habibi *et al.* 2017). The DGCCRF[6] indicates that, in 2015, 9 out of 10 French people declare that they have already carried out a collaborative consumption practice at least once, which makes the French the European consumers who have the largest appetite for the shared economy. As shown by the verbatim reports of users of carpooling or car-sharing platforms, the use of resources (car, petrol, roads, etc.) is optimized. These uses, which are based on the economy of functionality, can serve as a lever for the circular economy. Highlighting the relative financial (cheaper) or social (meeting new people, belonging to a community) benefits of these solutions promotes the circular economy, and works for the environment. In parallel with the many avenues proposed by ADEME to develop the circular economy, this other angle of attack is interesting: changing habits by insisting on two relative advantages other than ecology. The ecological conviction will come later or never, but in the meantime, the environmental impact will be real. The percolation of these uses of sharing economy will make it possible to cross Moore's chasm.

Table 8.1 suggests examples of communication aimed at consumers whose motivation is financial (the economic and strategic segments identified above by Siounandan *et al.*) or social.

	Current communication	**Communication with financial motivation**	**Communication with social motivation**
TooGoodToGo **https://toogoodtogo.fr/fr**	Fight against waste	Treat yourself to a cake for six people at the *Pâtisserie des rêves* by Philippe Conticini, one of the best pastry chefs for €16, which is 2.5 times cheaper than in a store!	Are you watching The Best Pastry Chef? Do you love Philippe Conticini? His sweet creations are now for you! Join the sweet tooth community that is enjoying itself!
Wag (We Act For Good) **https://www.wwf.fr/agir-au-quotidien/we-act-for-good**	Act on a daily basis	For the same price, consume organic or Label Rouge. All the merchants near you.	Take up challenges and participate in the consumer community! Join us by offering your recipes and tutorials! Come to our events near you!

6 https://www.economie.gouv.fr/dgccrf/Publications/Vie-pratique/Fiches-pratiques/economie-collaborative.

Phenix	Together, let's avoid waste	Get a basket of 3 kilos of fruit and vegetables for only €5! Eat beyond your means!	Get behind the scenes of your local merchants!
Frigo Magic **https://www.frigomagic.com/**	A quick and easy recipe idea to prepare an economical meal in your kitchen	Do you have two eggs and corn flakes left? Enjoy a magic tortilla. Easy recipes in 30 minutes max from what you have left in the fridge. How to enjoy and save!	Amaze your friends with easy recipes every day and join the community to share your favorites and gourmet tips!

Table 8.1. *Examples of new communication*

The managerial implications for companies, particularly start-ups, are twofold. On the one hand, on the substance, they would benefit from modifying their communication to highlight a relative advantage that may affect consumers who are not affected by militant discourse on waste. The relative benefits highlighted in this research are financial and social. On the other hand, in terms of communication channels, this communication should also extend beyond the company's website, application and social networks or blogs. It could be relayed in the media, particularly the women's press or morning television programs for housewives and pensioners as part of a press relations strategy. Lifestyle, cooking, travel or good tips influencers could promote it on their Instagram accounts or YouTube channels. These new prescribers would make it possible to reach a younger target group that abandons the traditional media (TV, press, radio, etc.) and places its trust in influencers. The implications for public authorities and associations supporting the circular economy also concern the promotion of new platforms and applications. A plan to help set up a press relations strategy, which is often costly for start-ups, would be useful. Finally, highlighting the other two relative advantages (financial, social) as a hook to change consumer behaviour, at least in the transition period, to cross Moore's chasm would be beneficial.

8.5. Conclusion

The sharing economy is exploding and now affects 90% of the French population. Airbnb, BlaBlaCar or Flickr are practices that are widely integrated into everyone's daily life, in all spheres of private life (travel, transport, food, etc.). On the other hand, the circular economy is struggling to develop and is stagnating with 13% of consumers engaged. The adoption of these disruptive practices cannot be based solely on ecological motivations. Two other motivations (financial, social) can help convince the rest of consumers to take the plunge, to adopt new behaviors that they often find complicated. By promoting a lower cost or an experience that is otherwise financially inaccessible, thrifty consumers and strategists, or 43% of the French population according to Siounandan *et al.*, can be sensitive to these discourses. Others will be more attracted by the social character, belonging to a community, the exchange of ideas or a playful experience that these new consumption patterns allow. Linking the circular economy to the driving force of the sharing economy seems a good idea, at least in the medium term. Finally, highlighting the fun nature of these applications with scavenger hunts, video tutorials, contests, events and customized solutions is also interesting for reaching the youngest. They are discouraged by the overly serious tone of the anti-waste discourse. Depending on age and economic situation, appropriate communication must be deployed. This opens up a wide range of possibilities for a hot topic of current and urgent interest.

An approach over a long period of several millennia makes it possible to be optimistic. In his 1923 essay on giving and the forms and reasons for exchange in archaic societies, sociologist Marcel Mauss argues that the market economy is very recent on a human scale. "It is our Western societies who have recently made man an 'economic animal'. But we are not yet all creatures of this genus. [...] *Homo oeconomicus* is not behind us, but lies ahead" (1968). The natural form is not the market exchange of the market society but exchange-giving, which inscribes the two parties in a gift-for-gift logic which implies not two individuals but a larger group (e.g. tribe). Economic phenomena cannot be limited to pure interest calculations but are total facts. They involve politics, symbolism and aesthetics, even religion. They induce rites, exchanges of politeness, symbolic goods, as illustrated by the potlach and kula. "The underlying reason for the exchange-giving is more about being than having." Will we see a paradigm shift with a return to these total exchanges far from pure utilitarianism? What if *Homo oeconomicus* is behind us?

8.6. References

Acquier, A., Daugigeos, T. and Pinkse, J. (2017). Promises and paradoxes of the sharing economy: An organizing framework. *Technological Forecasting & Social Change*, vol. 125, pp. 1–10.

Bardhi, F. and Eckhardt, G. (2012). Access based consumption: The case of car sharing. *Journal of Consumer Research*, vol. 39, pp. 881–898.

Belk, R. (1988). Possessions and the extended self. *Journal of Consumer Research*, vol. 15, pp. 139–168.

Belk, R. (2007). Why not share rather than own? *The Annals of the American Academy of Political and Social Science*, vol. 611, no. 1, pp. 126–140.

Belk, R. (2010). Sharing. *Journal of Consumer Research*, vol. 36, no. 5, pp. 715–734.

Belk, R. (2014a). Sharing versus pseudo-sharing in web 2.0. *The Anthropologist*, vol. 18, no. 1, pp. 7–23.

Belk, R. (2014b). You are what you can access: Sharing and collaborative consumption online. *Journal of Business Research*, vol. 67, pp. 1595–1600.

Bonet Fernandez, D., Petit, I. and Lancini, A. (2014). L'économie circulaire : quelles mesures de la performance économique, environnementale et sociale ? *Revue Française de Gestion Industrielle*, vol. 33, no. 4, pp. 1–25.

Bourg, D. and Buclet, N. (2005). L'économie de la fonctionnalité. *Futuribles*, no. 313, pp. 27–38.

Bradley, K. and Pargman, D. (2017). The sharing economy as the commons of the 21st century. *Cambridge Journal of Regions, Economy and Society*, vol. 10, no. 2, pp. 231–247.

Carboni, M. (2016). A new class of worker for the sharing economy. *Richmond J.Law Technol*, vol. 22, no. 4, p. 11.

Caroll, E. and Romano, J. (2011). *Your digital afterlife: When Facebook, Flickr and Twitter are your estate, what's your legacy?* New Riders, Berkeley.

Habibi, M.R., Davidson, A. and Laroche, M. (2017). What managers should know about the sharing economy. *Business Horizons*, vol. 60, no. 1, pp. 113–121.

Jevons, W.S. (1865). *The Coal Question: An Inquiry Concerning the Progress of the Nation, and the Probable Exhaustion of the Coal Mines.* MacMillan, London.

Maillefert, M. and Robert, I. (2014). Ecologie industrielle, économie de la fonctionnalité, entreprises et territoires : vers de nouveaux modèles productifs et organisationnels ? *Développement durable et territoires*, vol. 5, no. 1.

Mauss, M. (1968). *Essai sur le don : Forme et raison de l'échange dans les sociétés archaïques*, 4th edition. PUF, Paris.

Moore, G.A. (1991). *Crossing the Chasm.* Harper Business, New York.

Rogers, E. (2003). *Diffusion of Innovations*, 5th edition. The Free Press, New York.

Siounandan, N., Hébel, P. and Colin, J. (2013). Va-t-on vers une frugalité choisie ? *Cahier de recherche du CREDOC*, no. 302.

Tukker, A. (2004). Eight types of product-service system: eights ways to sustainability? Experiences from SusProNet. *Business Strategy and the Environment*, vol. 13, no. 4, pp. 246–260.

9

The Circular Economy from the Perspective of Voluntary Standardization

The circular economy covers a set of concepts whose definitions, modalities and impacts seem to differ from one country to another, from one company to another, from one consumer to another. Some wrongly consider it from the point of view of recycling only, others from the point of view of waste reduction, the pooling of resources or the eco-design of their products or services. In order to respond to this profusion of definitions and to help economic actors find their way around, the *Association Française de Normalisation* (AFNOR) published its first document in October 2018: the XP X30-901 standard.

AFNOR brings together experts, economic actors of all types (companies of all sectors, government representatives, local authorities, NGOs, consumer associations) to produce working tools, definitions and guides to facilitate exchanges between actors. Having a framework that is shared and approved by as many people as possible is a good way to move collectively towards more sustainable economic models.

9.1. Why and how has voluntary standardization appropriated the concept of the circular economy?

9.1.1. A need expressed by economic actors

9.1.1.1. *The interest of a standard*

The concept of a circular economy dates back some 30 years. It first appeared in 1990 in *Economics of Natural Resources and the Environment* (Pearce and Turner 1990), but it was in the 2000s that initiatives multiplied. Awareness of the increasing

Chapter written by Anne BENADY, Melodie MERENDA and Mérylle AUBRUN.

scarcity of our non-renewable natural resources (Meadows *et al.* 1970), climate change caused by the overexploitation of fossil fuels and the massive release of greenhouse gases into the atmosphere, mass production of waste and its impact on the environment (the 7th continent to name only one, "discovered" in 1997 by Charles Moore) are all elements that have motivated this profusion of initiatives: eco-design, Cradle to Cradle®, incentives for waste sorting and recycling, etc.

As recognized reference frameworks, voluntary standards are essentially intended to represent the consensual positioning of all actors (public or private) on a technical, organizational or methodological subject, etc. Clarifying and sharing became essential in view of the challenges raised by the circular economy. The standard and its mark – NF in France, EN in Europe and ISO internationally – ensure legitimacy that extends beyond borders and aims at a strong principle of continuous improvement. As early as 2016, the X30M[1] standardization commission, initially in charge of waste-related issues, initiated an inventory of existing standards in the field. It appeared that several subjects relating to the circular economy were already well "equipped" (this includes responsible purchasing, eco-design, environmental labeling), but that there was no framework to align all these approaches towards the same goal: reducing the impact on natural resources and effectively serving the three challenges of sustainable development: environmental, societal and economic.

Within this commission, a group of experts was created to prepare a first document: the XP X30-901 standard: "*Economie circulaire – Système de management de projet d'économie circulaire – Exigences et lignes directrices*". Among the experts who contributed to this document, we can mention: the Ministry of Ecological and Solidarity Transition, ADEME, the Suez Group, Tarkett, Italpollina, the Envie Federation, the European Institute for the Economics of Functionality and Cooperation, the Orée Association, the Institut National de l'Economie circulaire, etc.

The work was quick. The XP X30-901 standard was published in October 2018 after eight months of work. It has the particularity of being an experimental standard, which means that, given the extremely innovative nature of the subject it deals with, its lifetime is limited to three years. These three years are an opportunity to experiment "in the field" with the recommendations it makes. If the latter prove to be entirely relevant and beneficial for the actors, it will be approved as a French NF Standard.

1 Acronym referring to standards covering all environmental approaches.

XP X30-901 or how to demystify the codifications of normative documents

Standards have code names that may seem mysterious at first glance. Here is part of the mystery revealed:

XP = experimental standard. This means that it has not followed the complete process to become an official standard: it is not yet approved and therefore does not carry the prefix NF (for *norme française,* French Standard). It is therefore in a "trial period" for three years. At the end of these three years, the members of the commission that drafted it may decide to modify the document, renew its trial period or have it approved after having submitted it to a public inquiry.

In this case, as standardization work on the subject is ongoing at ISO (International Standards Organization), it is possible that the initial document will be replaced by a document with the prefix NF ISO.

X30 = the code name of the commission that produced the document (its full code is X30M). More specifically, this group of experts is part of a larger committee working on environmental management and social responsibility.

901 = the document code, among all the normative documents produced by the X30M commission. The 1 at the end of the code means that it is a management system standard, such as ISO 9001 (quality management system), ISO 14001 (environmental management system) or ISO 37001 (management system for sustainable development within territorial communities).

Box 9.1. *Codifications demystified*

9.1.1.2. *An operational standard*

Nearly 50 actors contributed to the work, each with its own vision, its own specific challenge, but the same motivation. This motivation was based on common definitions but, above all, identified the need to build a practical and operational translation of the concept to support and guide all those wishing to implement this paradigm shift. The aim was not so much to make a certified standard as a standard framed by a management system that could easily be integrated into a conventional system. The maturity of the approach can thus be validated thanks to the requirements of the reference framework but above all to the measurement of the achievement of objectives in the monitoring of the project.

To achieve this, several principles guided the discussions:

– exploit all the wealth of initiatives implemented by public and private actors already engaged in the transition of their economic model and go beyond the simple "collection of good examples";

– start from concrete projects to propose an operational project management tool;

– take into account the entire value chain to build this common framework for dialogue between stakeholders;

– be part of a sustainable development perspective.

9.1.2. *Towards an ISO standard*

France is obviously not the only country to identify that a change in economic model is inevitable. In 2015, the European Commission adopted an action plan to accelerate the European transition to the circular economy (54 measures) and in June 2018, the circular economy package was published in JOUE. As for China, becoming aware of its rapid development and the resulting consequences on the environment, it has been able to invest in projects aimed at reducing greenhouse gas emissions, creating eco-parks and eco-cities, etc., since 2013. A law promoting the circular economy proposes, even in 2013, more than 80 macroeconomic monitoring indicators for the circular economy.

AFNOR therefore proposed to take this subject to the international level, which was approved by more than 40 countries. A technical committee, TC 323[2], has been created within the ISO (International Standards Organization) to produce not one, but many reference documents on the subject of the circular economy. Indeed, the needs are multiple: principles, reference framework, terminology, management system, evaluation tools and indicators, territorial approach, etc., all subjects on which the members of this new technical committee will work in the coming years.

TC 323 met for the first time in May 2019 to establish the roadmap for the work. Four working groups have been defined and bridges established with other ISO committees to ensure that all the work supports the transition to a circular economy. Among these committees, we can mention:

– quality management and quality assurance (ISO TC 176);

– environmental management, and, in particular, the subcommittee 5 which works on lifecycle assessment (ISO TC 207);

2 TC for Technical Committee.

– sustainable cities and communities (ISO TC 268);

– economics of sharing (ISO TC 324).

To date, TC 323 members are present on all five continents. More specifically, 55 countries are participating and 13 countries are observer[3] members. Both developing and developed countries are represented on the committee. The participation of so many countries reflects an awareness of the urgency of a transition to a circular economy.

Through its proactive action within ISO, and its desire to defend a clearly defined vision of the circular economy for sustainable development, AFNOR has enabled France to establish its leadership and influence on the international scene.

9.2. The main principles of the XP X30-901 standard

The experimental standard, "XP X30-901 – Circular economy project management system" is one of the operational tools for conducting any type of circular economy project. It clearly positions the circular economy as part of a sustainable development perspective.

The circular economy is defined as: "An economic system of trade and production which, at all stages of the lifecycle of products (goods and services), aims to increase the efficiency of resource use, reduce the impact on the environment while allowing the well-being of individuals, where the value of products, materials and resources is maintained in the economy as long as possible and waste production is reduced to a minimum."

The circular economy therefore leads to an interest in the entire lifecycle of the product or service, and all impacts and/or transfers of impacts between seven identified areas of action:

– sustainable procurement;

– eco-design;

– industrial symbiosis;

– economy of functionality;

– responsible consumption;

3 Observer members participate in meetings but do not have the right to vote. They are in a standby position but can become an active member depending on the progress of the work.

– extension of the duration of use;

– effective management of end-of-life materials or products.

The notion of a circular economy project refers both to an activity (providing products or services) and to all the activities of an organization. The standard, although a management system standard, can also be applied to the construction or reinforcement of the strategy of the company that has chosen to invest in the circular economy. The objective is to develop the traditional model in order to make it more efficient and with as little environmental and social impact as possible. Some examples of projects include: redefining a company's business strategy, modifying a logistics, design or procurement process, etc.

Based on the High Level Structure (HLS) – the universal structure of management system standards – this standard follows the principle of continuous improvement.

It is divided into 10 sections and has the particularity of containing requirements, in other words, action and compliance obligations as well as recommendations to help understanding and implementation. The existence of requirements thus makes it possible to declare conformity with the standard if they are fully met. But the main objective is to enable each organization to apply the challenges of the circular economy to its activity or activities.

9.2.1. *Structure and content of the reference framework*

9.2.1.1. *The context of the organization and the 3X7 matrix*

A standard always applies to an organizational context, especially in the area of sustainable development. Clarifying the organization's context therefore consists of defining the scope of application of the standard, by taking into account the company's sustainable development issues in terms of both activities and geographical locations. We will discuss internal and external issues.

Diagnostic methods such as the one proposed in the standard on social responsibility (NF X30-029) can be used to identify and prioritize these issues.

The need to meet the challenges related to sustainable development and the seven defined areas of action allows the organization to identify the objectives and goals necessary for the implementation of a circular economy project. The standard proposes a so-called "3X7" matrix for this purpose, shown in Figure 9.1. It also provides some examples of its use. A real project management tool, the 3X7 questioning matrix is a methodology for identifying the priority actions to be implemented. It also allows a follow-up over time of the project which contributes to its continuous improvement.

While the standard requires systematic consideration of the three dimensions of sustainable development and its possible transfers of impacts, it recommends the use of the 3X7 matrix to achieve this. We therefore have a combination of purpose requirements and a methodological recommendation. However, the use of the matrix is required when building its objectives and associated action plan.

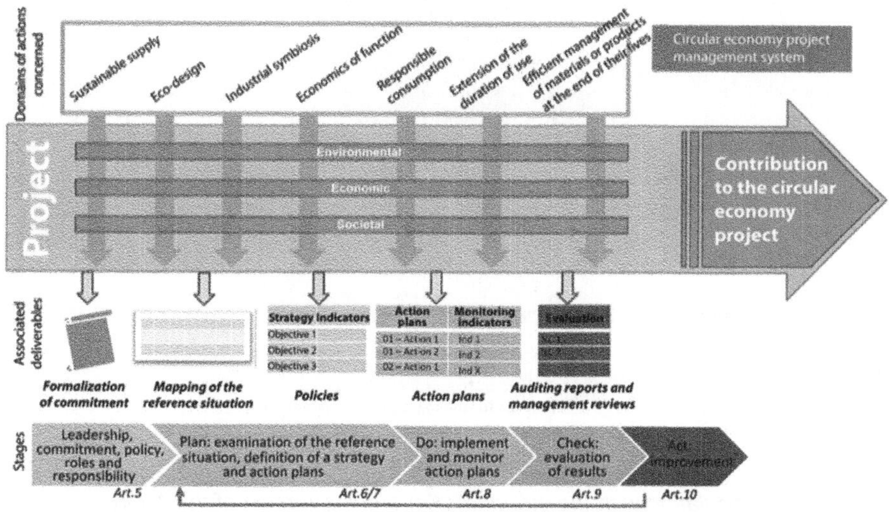

Figure 9.1. *The 3X7 matrix, a schematic representation of the project management system contributing to a circular economy (extracted from the XP X30-901 standard). For a color version of this figure, see iste.co.uk/delchet/circular.zip*

9.2.1.2. *The place of leadership, politics and management system in circular economy project management*

Like all management system standards, it is necessary to build your circular economy project on the strong involvement of the organization's management. This involvement can be measured by taking into account the written policy shared by all, the objectives defined according to the challenges, the context and the methodological analysis specific to the circular economy, but also according to the human, technical and financial resources made available. All objectives must be able to be integrated into the various existing processes and comply with the principle of continuous improvement. The circular economy management system put in place will therefore naturally complement the systems already implemented within the company.

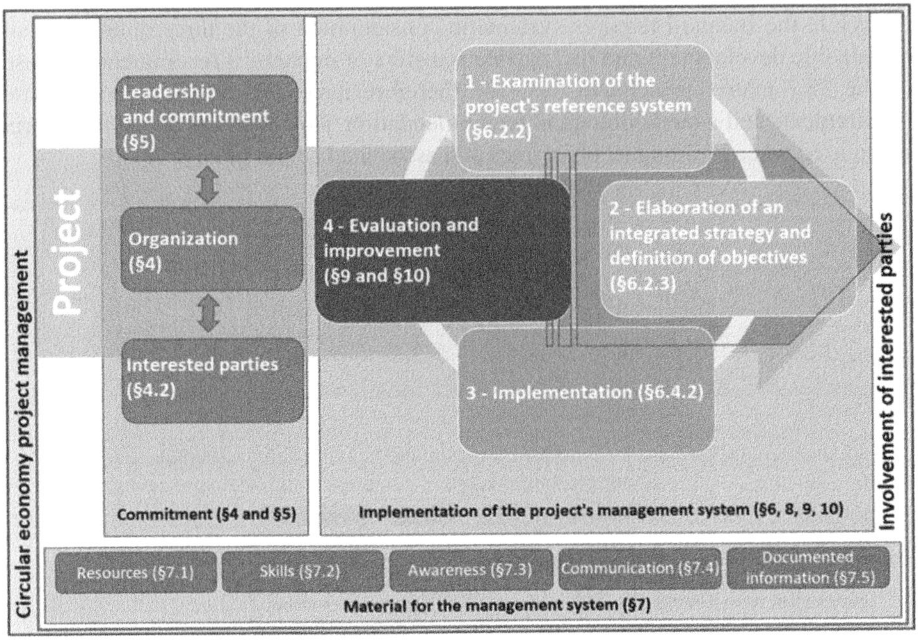

Figure 9.2. *Schematic representation of the project management system and its articulation with the standard. For a color version of this figure, see iste.co.uk/delchet/circular.zip*

To optimize the choice of priority objectives and targets, a diagnosis characterizing the organization's initial situation, in relation to the circular economy project, is required. Based on the 3X7 matrix, the latter will also integrate the regulatory requirements applicable to the organization, such as the constraints related to the signing of a Regional Natural Park charter, for example, or those related to ICPE regulations.

Compliance with sustainable development principles, such as transparency and accountability, should enable stakeholders to be involved in the development of strategy and objectives. This standard therefore positions itself at the interface between the organization's strategy and an approach that may initially prove to be experimental on a product, project and/or company service related to the circular economy project.

To monitor the circular economy project identified by the company, the latter will have to set up a series of key indicators to measure the effectiveness of the approach and its contribution to the seven areas of the circular economy, avoid the transfer of possible impacts, and guarantee the principle of continuous improvement. From an organizational point of view, the Management Review is an opportunity to review at regular intervals, in order to identify opportunities for continuous improvement.

Each step of the project and its follow-up are documented and must be able to be used as evidence of compliance with the requirements of the experimental standard.

9.3. The circular economy project management standard, a suitable tool for evaluating, enhancing and improving projects

9.3.1. *Towards a "Circular Economic Assessment"*

To support organizations in their circular economy project, AFNOR Certification has developed "*L'évaluation économie circulaire*", a complementary tool to the standard that evaluates the integration of its recommendations. It is an effective tool for changing production and consumption patterns.

Based on the 3X7 matrix, the circular economic assessment allows organizations (regardless of size), business sectors and legal structures to (re)question their actions and responsibilities:

– on their value chain upstream of product production (sustainable supply);

– within their production process, manufacturing, R&D and support functions (eco-design, extension of the service life, responsible consumption);

– on the supply side (economy of functionality, reparability, reuse);

– at the level of the territory and exchanges with local economic and non-economic actors (industrial and territorial ecology);

– at the end of the life of the product placed on the market, in order to ensure the health and safety of all consumers (effective management of materials or products at the end of their life).

The benefits of a circular economic evaluation based on the experimental standard are transversal:

– prioritization of circular economy actions in order to identify urgent actions to be carried out internally as well as with external stakeholders, within the company's territory;

– optimization of the control of economic, social, societal and environmental risks;

– improvement of the control of the project development process;

– development of the capacity for anticipation and resilience;

– emergence of new modes of consumption and production, even new economic models (functional economy, social and solidarity economy, collaborative economy, etc.);

– better acceptance of activities, especially industrial activities;

– strengthening the mobilization of internal stakeholders around company projects (from employees to members of the management committee);

– improving consistency and convergence between organizational management system standards (ISO 9001 on quality, ISO 14001 on environment, ISO 50001 on energy, ISO 45001 on occupational health and safety) and standards of guidelines, and good practice for sustainable development (ISO 26000 on social responsibility, ISO 20400 on responsible purchasing, ISO 37101 on sustainable development in the territories, etc.);

– continuous improvement of actions and objectives.

9.3.2. *First feedback from users*

The first companies evaluated by AFNOR Certification on the basis of the experimental standard appreciate its adequacy with the QSE[4] management systems (potentially) already in place: indeed, having an integrated management system promotes the planning, implementation and completion of circular economy projects.

However, how did these companies identify the project to be evaluated?

To date, two approaches can be distinguished: on the one hand, a global approach, for companies that consider that the circular economy is already their core business and that have the company's global strategy evaluated. On the other hand, a more iterative approach, starting from a more "modest" product or range development project, leading to the involvement of all the actors of the organization around this development (purchasing, conception/R&D, marketing, customers or users, etc.).

Finally, the evaluation of the circular economy on the basis of the standard makes it possible to challenge one's circular economy project against the requirements of management systems and guidelines, which is then more in line with a project maturity evaluation perspective.

4 QSE: Quality, Safety and Environment, also known as Integrated.

9.4. References

Meadows, D.H., Meadows, D.L., Randers, J. and Behrens, W.W. (eds) (1972). *Halte à la croissance?* Fayard, Paris.

Pearce, D.W. and Turner, R.K. (1989). *Economics of Natural Resources and the Environment.* Johns Hopkins University Press, London.

ISO 37101 (2016). Développement durable au sein des communautés territoriales — Système de management pour le développement durable — Exigences et lignes directrices pour son utilisation. ISO.

XP X30-901 (2018). Economie circulaire — système de management de projet d'économie circulaire — Exigences et lignes directrices. AFNOR

PART 3

Activity Sectors through the Prism of the Circular Economy

10

Circular Economy and Construction

10.1. Introduction

This chapter focuses on the application of the concept of a circular economy to construction. The latter is understood in the sense of an economic sector and, according to Vincent (1985), comprises five main segments: building materials, distribution-trading, construction and public works, designers and project managers, and contracting authorities. To these five segments, we think it is appropriate to add construction and demolition waste managers.

This chapter consists of four sections. The environmental issues related to construction are first briefly presented. Then, sixteen elements of definition of a circular economy applied to construction that were found in the literature are studied. In the third section, a brief overview of policies, research and development projects and construction and urban development projects are provided. Finally, four main limitations to these definitions, policies and projects are discussed.

10.2. Global environmental issues related to construction

Construction generates significant material flows from the extraction of natural resources, the production of materials and their use on construction sites, to waste management. Construction materials are the first materials consumed by humankind after water (IRP 2018). Among construction materials, non-metallic minerals and in particular aggregates (sand and gravel) are the most commonly used (Krausmann *et al.* 2009). Aggregates are used in concrete, asphalt mixes and road pavement sub-bases. They are also used as backfill.

Chapter written by Vincent AUGISEAU.

The construction, maintenance, repair and demolition of buildings and networks, as well as certain civil engineering works such as site leveling, generate outgoing flows of materials commonly known as construction and demolition waste. The latter constitute by their mass the first solid waste generated by humankind (Krausmann *et al.* 2017). They are composed of non-metallic minerals from concrete, stone or brick structures, as well as metals, wood, plastics and bituminous materials. They also consist of excavated materials, i.e. soil and minerals removed from a site during excavations.

On a global scale, only one-third of the mass of waste excluding excavated materials is recycled as construction materials (*ibid.*). Recycled materials represent only one-tenth of all the construction materials consumed (*ibid.*). Moreover, they are mainly used for purposes that are unsuited to their quality. For instance, recycled aggregates are predominantly used as backfill and for road construction, uses which are the least demanding in terms of quality (see Hashimoto *et al.* 2007; Augiseau 2017).

Material and waste flows are a source of environmental impacts and land-use conflicts. The production of materials generates a significant extraction of natural resources. The latter are largely non-renewable and sometimes in a situation of scarcity: on a global scale, such as copper (Gordon *et al.* 2006), or on a local scale, such as sand (Peduzzi 2014). In addition, mining activities, as well as the expansion of the built environment and the management of construction and demolition waste, generate land use that temporarily or permanently reduces the possibility of producing or extracting both renewable and non-renewable resources (Bringezu 2002).

In addition to contributing to the depletion of non-renewable resources, construction generates pressures on natural environments and harmful effects on human health. Mineral extraction transforms landscapes and impacts aquatic environments, fauna and flora. This is particularly the case for the extraction of marine aggregates, which generates conflicts of use with fishing activity, as well as risks in terms of biodiversity, fisheries resources and coastal erosion (Peduzzi 2014).

The production of materials also generates emissions into the air. Producing cement generates 5 to 6% of the world's anthropogenic greenhouse gas emissions (Mishra and Siddiqui 2014). In addition, on a local scale, air emissions from some cement plants cause respiratory diseases (*ibid.*). In addition, energy flows that are also sources of environmental impacts are associated with material flows generated by construction (Pullen 2000).

World population growth combined with an increase in the proportion of the population living in cities will lead to the development of built-up areas, which will accentuate the environmental challenges associated with construction. According to Elmqvist *et al.* (2013), 60% of the built space that will be needed by the world's population in 2050 has not yet been built. Krausmann *et al.* (2017) estimate that the mass of materials accumulated in buildings and networks will quadruple by 2050. Global material consumption, which increased 10-fold from 1950 to 2005 (Krausmann *et al.* 2009), could therefore double by 2060 compared to 2011 (OECD 2018). The generation of construction site waste, which increased sixfold from 1950 to 2005, will also continue to grow (Krausmann *et al.* 2017).

Figure 10.1 schematically shows the material, energy and emission flows generated by the construction industry, as well as the environmental impacts and land-use conflicts. According to Krausmann *et al.* (2017), approximately one-third of construction and demolition waste excluding excavated materials is considered to be recycled for construction, regardless of the form of recycling. In the absence of an estimation source, excavated materials and emissions flows are not represented in proportion to other materials flows.

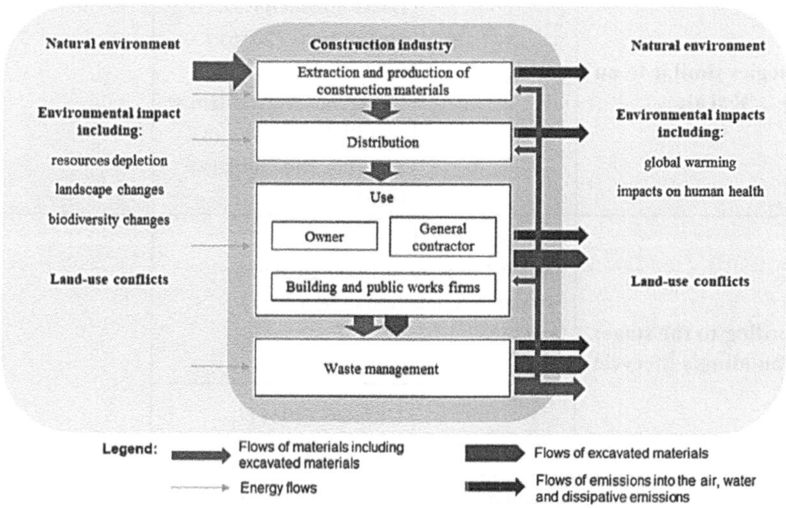

Figure 10.1. *Schematic representation of flows, environmental impacts and land-use conflicts generated by the construction industry, For a color version of this figure, see iste.co.uk/delchet/circular.zip*

10.3. Sixteen elements of definition

Applying the concept of a circular economy to construction could be a response to the environmental issues presented above. Economic and social benefits could also result from this (European Commission 2019). However, the concept of a circular economy applied to construction is not currently the subject of a consensual and shared definition. Thus, the different existing definitions are studied here from a review of technical and scientific literature. Thirteen documents presenting sixteen elements of definition are selected.

<table>
<tr><th></th><th>Construction</th><th>Building</th><th>Built environment</th></tr>
<tr><td>Three principles of a circular economy according to the Ellen MacArthur Foundation</td><td colspan="2">Athanassiadis (2017) (strategies)</td><td>EMF (2017) (principles)</td></tr>
<tr><td>Six levers for a transition to the circular economy according to the Ellen MacArthur Foundation</td><td colspan="2"></td><td>ARUP (2016) (actions or elements);
BAM et al. (2016) (elements)</td></tr>
<tr><td>Strategies similar to an R scale</td><td></td><td>Circle Economy et al. (2018) (principles);
Geldermans (2016) (steps);
Gemeente Amsterdam et al. (2019) (principles);
Sommer and Guldager (2016)</td><td></td></tr>
<tr><td rowspan="2">According to the stages of a building's lifecycle</td><td colspan="2">Adams et al. (2017) (aspects);
French Ministry of Ecological and Solidarity Transition et al. (2017) (levers);
CSTC (2017) (axes)</td><td rowspan="2"></td></tr>
<tr><td></td><td>UKGBC (2019) (principles)</td></tr>
<tr><td>Definitions close to that of sustainable development</td><td>Circular Economy Programme cited by Circle Economy et al. (2018)</td><td>Circle Economy et al. (2018) (impact areas);
Gemeente Amsterdam et al. (2019) (characteristics or themes)</td><td>Pomponi and Moncaster (2017) (dimensions)</td></tr>
</table>

Table 10.1. *Sixteen elements of the definition a of circular economy applied to construction, building and the built environment, classified into five groups*

The scope of the review does not fully cover the construction sector as defined in the introduction. Indeed, some documents adopt an object-based approach and deal with the building or built environment[1]. However, they provide definitions that are relevant to our study.

Five main sets of definitions, presented in Table 10.1, can be distinguished. Several documents, although they refer to the construction sector or industry, are located between construction and building. In addition, definitions are formulated in a wide variety of forms. Also, the terms used by the authors are specified in brackets in the table.

10.3.1. *Three principles of the circular economy according to the Ellen MacArthur Foundation*

Three principles of a circular economy have been proposed by the Ellen MacArthur Foundation (EMF): "preserve and enhance natural capital by controlling finite stocks and balancing renewable resource flows"; "optimize resource yields by circulating products, components, and materials in use at the highest utility at all times in both technical and biological cycles" and "Foster system effectiveness by revealing and designing out negative externalities" (EMF 2015, p. 23).

Two documents use these principles. First, a definition of the circular-built environment based on these principles is proposed by EMF (2017, p. 7):

> "A built environment that is designed in a modular and flexible manner, sourcing healthy materials that improve the life quality of the residents, and minimize the use of virgin materials. It will be built using efficient construction techniques, and will be highly utilized thanks to shared, flexible and modular office spaces and housing. Components of buildings will be maintained and renewed when needed."

In addition, the strategies constituting the model for the construction sector within the Brussels region are derived from the three EMF principles (Athanassiadis 2017). The construction sector defined by the author excludes the construction and demolition of infrastructure. The model highlights several sets of loops: "care, repair, maintenance; reuse, repurposing; upcycling, remanufacturing, reconditioning; recycling, composting" (*ibid.*, p. 38, author's translation). The author recommends "promoting the tightest loops of the model [as] recycling should be minimized and the elimination of building materials [...] should be prohibited" (*ibid.*, p. 37, author's translation). The current model of the construction sector in Brussels is compared with the theoretical model developed by the EMF "in order to understand

1 The built environment, which includes buildings and networks, is an intermediate scale between the building and the city.

the transition needed to move from one to the other" (*ibid.*, p. 36, author's translation).

10.3.2. *Definitions from the six levers for a transition to a circular economy according to the Ellen MacArthur Foundation*

In addition to the three principles presented above, the EMF (2015, p. 26) defines a framework composed of six levers which "offer companies a tool for generating circular strategies and growth initiatives". These levers are formulated in the form of verbs constituting the acronym RESOLVE: REgenerate, Share, Optimize, Loop, Virtualize and Exchange (*ibid.*).

Two documents are based on these levers. A census of projects representative of a circular economy for the built environment published by the EMF (BAM *et al.* 2016) presents examples of actions for each of the levers. The first three include regenerating and restoring soils by building on wasteland; sharing housing, offices or infrastructure; and optimizing construction through prefabrication, optimizing the use of space through the use of vacant plots and dense urban growth. The next three measures involve in particular optimizing the end of life of buildings and materials, modularity of buildings and refurbishment of materials, virtualization of products and processes and, finally, an exchange using more efficient materials and technologies.

A second document is based on the levers to illustrate how a circular economy can be applied to the built environment (ARUP 2016). It also cross-references the levers with an adaptation of the concept of building in "layers", designed by architect Frank Duffy (1990) and developed by Stewart Brand (1994). According to this model, buildings are made up of six separate and interconnected layers, each with a different life span. According to the authors, the application of this concept would allow a better separation and removal of each element and thus facilitate reuse and recycling. It would also allow each element to be repaired, replaced, changed or adapted without affecting the building or infrastructure as a whole, thereby reducing obsolescence and increasing flexibility and service life.

To the six formulated layers defined by Brand (1994) is added a so-called system layer with the longest life span which, according to the authors, allows them to apply their proposed circular economy approach to a neighborhood or city. In order of maximum lifetime, the seven layers are: system (urban structures and services including transport, energy and water networks); structure (foundation and load-bearing elements); skin (façade and exterior); site (location of the building); services (pipes, wires, energy and heating systems); space (walls and floors); and miscellaneous stuff (furniture, lighting and information and communication technologies) (ARUP 2016). Actions for the 42 elements formed by the crossing of the six levers and seven layers are proposed.

10.3.3. *Definitions according to a strategy similar to an R scale*

According to Reike *et al.* (2018, pp. 249–250), "in [a circular economy], the distinction of various preservation stages of resource value using hierarchical 'R-ladders' or imperatives, is an essential operationalization principle." A variety of R scales are identified by the authors, the most shared scale being reduce, reuse and recycle. Three documents on the application of the concept of a circular economy to buildings present strategies similar to an R scale.

Some Dutch authors proposed four building design strategies: "reduce, synergize, supply and manage" (Circle Economy *et al.* 2018, p. 15). These strategies are prioritized. The aim is to reduce the demand for resources, then when the demand for resources and the associated impacts have been reduced to a minimum, one identifies local synergies that can satisfy these demands. When the opportunities for sourcing from synergies have been exploited, requests are covered by clean, renewable, recycled or low-impact sources.

A fairly similar scale of imperatives is proposed by Geldermans (2016). The author uses the New Stepped Strategy, defined by van den Dobbelsteen (2008) based on Cradle to Cradle concepts, which consists of reducing demand, reusing and recycling and then meeting the remaining demand in a sustainable way by only releasing emissions with no environmental impact. Geldermans (2016) proposes a six-step strategy. In the first step, "the added value of the intended functions and their materialization" is evaluated, which means the need for a new construction is questioned. Step 2 aims to explore "current and future vacant buildings with regard to availability and usability" (*ibid.*). Step 3 is about integrating "change in a new adaptable design", and step 4 is about applying "intelligent dimensioning" (*ibid.*). The two last steps consist of exploring "the availability and usability of existing materials" and integrating "high-quality future reuse" (*ibid.*).

Finally, although it is only partially similar to an R scale, the definition of the circular building proposed by Sommer and Guldager (2016) highlights the requirements related to the reuse and recycling of waste. Indeed, according to the authors, "a circular building is a temporary aggregation of components, elements, and materials with a documented identity, recording their origin and possible future repurposing, assembled in a certain form, which accommodates a function for an established period of time" (*ibid.*, p. 132).

10.3.4. *Definitions according to the stages of a building's lifecycle*

Four documents propose elements for defining the application of a circular economy to construction or building using a lifecycle approach. CSTC (2017) proposes an "approach" and "concrete themes of the circular economy for the construction sector" (*ibid.*, p. 6, author's translation) structured around three axes:

> "The adaptability of construction, the choice of materials and assemblies as well as the minimization of waste production [...]; at the end of life, selective deconstruction, reuse, remanufacturing and recycling [...]; new economic models of the circular economy such as lifetime extension, sharing economy and functionality economy." (*ibid.*, p. 108, author's translation)

The document also refers to Stuart's (1994) layers model.

A second document proposes a "framework for defining a circular economy [...] for buildings [...] in order to support the actors and facilitate its operational implementation in the construction sector" (Ministère de la transition écologique et solidaire *et al.* 2017, p. 6, author's translation). According to the authors, the framework makes "the link with the reference framework for sustainable building" (*ibid.*, author's translation) that they have previously formulated. This framework consists of 15 levers grouped into five groups: territorial optimization of flows, sobriety, lifetime extension, resources production so as to limit waste and stakeholder management. In this context, some pillars of the circular economy, according to ADEME (2014), can be found: sustainable procurement, eco-design, functional economy, lifetime extension, recycling.

An article by researchers at Loughborough University in the UK provides a list of key aspects for the application of a circular economy over the lifetime of a building (Adams *et al.* 2017): design, manufacture and supply, construction, in use and refurbishment and end of life. At the design stage, some concepts already mentioned above are indicated: design for disassembly, design for adaptability and flexibility, design for standardization, designing out waste, designing in modularity (*ibid.*). For all stages, the authors recommend the management of information.

A guide published by UKGBC (2019) defines five sets of principles partly organized around the stages of a building's lifecycle. These principles are: reuse; design buildings for optimization; standardization or modularization; servitization and leasing; and responsible design and construction. The actions proposed for the application of these principles are similar to the aspects of Adams *et al.* (2017). In addition, as well as ARUP (2016), the guide refers to Stewart Brand's (1994) layers building model and proposes strategies for each layer.

10.3.5. *Definitions of the circular economy close to that of sustainable development*

Two sets of documents propose definitions of a circular economy applied to the built environment or building close to definitions of sustainable development. Pomponi and Moncaster (2017), in an article which aims to define a framework for research on the circular economy applied to the built environment, propose six dimensions for the latter: economic, environmental, technological, societal, governmental and behavioral. The authors refer to the three pillars of sustainable development. They also propose a definition of a circular building: "a building that is designed, planned, built, operated, maintained, and deconstructed in a manner consistent with [circular economy] principles" (*ibid.*, p. 711). If these principles are not clearly explained, the authors refer to the six dimensions mentioned above.

Circle Economy *et al.* (2018) also propose elements of a definition similar to that of sustainable development. The latter are based on the Circular Economy Programme defined by the Dutch government, which includes the objective of:

> "the development, use and reuse of buildings, areas and infrastructure, without avoidable depletion of natural resources, pollution of the environment or negatively impacting ecosystems. Construction which is economically responsible and contributes to the wellbeing of humans and animals, now and in the future." (cited in Circle Economy *et al.* 2018, p. 11)

The authors also refer to the work of the EMF, as well as Gladek (2017), which "considers the circular economy from a broader sustainability perspective" (Circle Economy *et al.* 2018, p. 8). An adaptation of the Dutch government's definition for a building is proposed:

> "A building that is developed, used and reused without unnecessary resource depletion, environmental pollution and ecosystem degradation. It is constructed in an economically responsible way and contributes to the wellbeing of people and other inhabitants of this Earth. Here and there, now and later. Technical elements are demountable and reusable, and biological elements can also be brought back into the biological cycle." (*ibid.*, pp. 11–12)

Seven impact areas are defined: materials and energy; water; biodiversity and ecosystems; human culture and society; health and well-being; and multiple forms of value (*ibid.*). These areas are cross-referenced with the four building design strategies presented above. Similar elements of definition are formulated by some of the authors in a document for the city of Amsterdam (Gemeente Amsterdam *et al.* 2019).

10.3.6. *Cross-referenced analysis of definitions*

Common points and differences can be observed between the elements of definition. First of all, we can note some common references to a large number of the documents: the texts published by the EMF and the concepts of Cradle to Cradle from McDonough and Braungart (2002), which themselves strongly influenced EMF. At the building level, Brand's (1994) layers model is frequently used, as are the concepts of design for disassembly, design for adaptability or flexibility or design in modularity. All the definitions emphasize, in a more or less marked (pronounced) and exclusive way, the importance of recycling construction and demolition waste, generally by anticipating future waste flows at the design stage.

The review shows that few documents cover all segments of the construction industry as defined in the introduction. Indeed, a majority of studies focus on buildings and, in particular, the new construction of buildings, the only modality taken into account by some authors when considering the application of the concept of the circular economy. In addition, the definitions use very different terms. Most of them define principles or strategies, as well as aspects, axes, characteristics, dimensions, impact areas, elements, stages, levers or themes. The majority of authors favor the means to be used to achieve a circular economy, while others favor the purpose of the latter.

The recommendations are generally not prioritized. This is the case for the three principles and six levers proposed by the EMF. However, several authors formulate hierarchical strategies that highlight the importance of reducing materials and waste flows and their associated environmental impacts. Finally, we can observe more or less broad thematic scopes, some authors focusing only on the material aspect of the circular economy, while others integrate all or part of the themes usually related to sustainable development. Before returning to some of the analytical points briefly introduced here, we propose to draw up a brief table of policies and projects.

10.4. Policies and projects aiming to apply the concept of the circular economy to construction

10.4.1. *Policies*

While the variety of definitions that can be observed seems to reflect a lack of stability in the concept of the circular economy applied to construction, the latter is nevertheless the subject of policies and projects: European, national or local policies; research and development projects leading to the creation of tools; and construction and urban development projects.

Construction and demolition is one of the five priority sectors defined by the European Commission for the European Action Plan for a Circular Economy and for which the Union "should [...] continue to support research, innovation and investment" (European Commission 2019, p. 12). The same document identifies the built environment as one of the "many other sectors with high environmental impact and strong potential for circularity" (*ibid.*).

Several European Union member countries have defined circular economy policies for construction. This is particularly the case in the Netherlands, where the definition from the Circular Economy Programme was presented earlier. This is also the case in France where several measures of the roadmap towards a circular economy published in 2018 deal with the construction sector and aim in particular at promoting the recycling of construction and demolition waste (*Ministère de la transition écologique et solidaire* and *Ministère de l'économie et des finances* 2018).

Construction is also a key sector for 11 of the 12 strategies for the circular economy supported by cities observed by the European Economic and Social Committee (2019). Of the 210 local circular economy strategies studied by Petit-Boix and Leipold (2018), 9% focus on the recycling of construction and demolition waste or the eco-design of buildings. As well as the circular economy policies for construction initiated by the city of Amsterdam and the Brussels region mentioned above, approaches are being developed in France. This is particularly the case for the roadmap towards a circular economy of Paris, the first three actions of which concern construction and urban development (Mairie de Paris 2017).

10.4.2. *Research and development projects*

The application of the concept of the circular economy to construction has led to research and development projects in Europe and France, resulting in the creation of tools. Level(s), "the first framework of indicators for measuring sustainability in the sector" according to the European Commission (2019, p. 9), is being tested by 130 projects in Europe in 2019. The second objective is to "optimize the building design, engineering and form in order to support lean and circular flows, extend long-term material utility and reduce significant environmental impacts" (European Commission Joint Research Centre 2017, p. 10).

The European research project Horizon 2020 Buildings as Material Banks (BAMB) carried out from 2015 to 2019 led to the development of the evaluation tool, the Circular Building Assessment (Hobbs 2019). The latter aims to evaluate the difference between a standard building design scenario and a so-called circular scenario. It is based in particular on Brand's (1994) layers model. A relatively similar tool, the CE Meter, was developed by Geraedts and Prins

(2015). Leising *et al.* (2018) propose a tool to facilitate the organization of a project and in particular to enhance collaboration in the building sector.

Two calls for research and development projects from ADEME in 2012 and 2014 focus on circular economy for construction. The vast majority of these projects aimed at developing the recycling of construction and demolition waste, with only two of the 18 projects dealing with waste prevention. The OVALEC project, financed by ADEME under another call for projects, resulted in the development of an evaluation tool composed of four sets of indicators about the supply and use of aggregates and waste management (CSTB *et al.* 2018). Four so-called circularity indicators have also been defined in order to integrate criteria related to the circular economy into *Haute Qualité Environmentale* (HQR – a standard for green building) certification (Oury 2019). This work on indicators is being pursued as part of a project launched in 2018 by the Fondation Bâtiment Energie (2018).

10.4.3. *Construction and urban development projects*

Examples of projects illustrating the application of the concept of circular economy to construction are provided in the documents reviewed. ARUP (2016) presents 41 case studies that "exemplify elements of circularity [while] few of them are perfectly circular" (*ibid.*, p. 18). Indeed, according to the authors, "circular practices tend to occur at the individual component or asset level. They include modular, prefabricated and off-site construction, design for disassembly, materials reuse and recycling, and designing out waste" (*ibid.*, p. 43). Therefore, "the prospect of linking all aspects of the built environment through a fully inclusive and comprehensive circular economy remains a challenge" (*ibid.*). BAM *et al.* (2016) present 12 examples including research projects such as BAMB.

Two documents by van den Dobbelsteen (2008) and Leising *et al.* (2018) present the example of the Park20|20 business park in Hoofddorp in the Netherlands. The concept of Cradle to Cradle was used in the design of this neighborhood by McDonough and his associates. Six innovative business models are identified by Leising *et al.* (2018, p. 981): "Create value from waste [...]; Deliver functionality without ownership [...]; Optimize material efficiency [...]; Substitute with renewables [...]; Repurpose for society by designing buildings with a healthy indoor climate [...]; Inclusive value creation, via alternative solutions for ownership."

10.5. Four main limitations

In our view, the definitions given to the concept of the circular economy applied to construction and the policies and projects aiming at this application have four main limitations: an uncertain delimitation of the scope of the circular economy, a low consideration of the territorial context, a scale for the application of the circular economy concept that is often too narrow, and an insufficient articulation between the circular economy and spatial planning.

10.5.1. *An uncertain delimitation of the scope of the circular economy between waste management and sustainable development*

The first limit that can be identified concerns the extent of the scope of the concept of the circular economy applied to construction. A focus on the recycling of construction and demolition waste, including reuse, can be observed. This focus is even more pronounced in policies and projects, particularly in research and development. Emphasis on the recycling of construction and demolition waste is entirely relevant given that the issues related to this waste have only recently been addressed, in particular by strengthening regulatory objectives. However, the risk is to lock the circular economy into a curative or end-of-pipe approach and to reproduce some of the limitations identified for the application of the concept of industrial ecology (O'Rourke *et al.* 1996). Narrowing the circular economy by relating it to waste management only is a limit observed by Ghisellini *et al.* (2016) for activities other than construction.

Moreover, it seems to us that this risk is more pronounced when the principles or strategies defined for the application of a circular economy to construction (and sometimes for all activities) are not prioritized. This is particularly the case for the three principles and six levers in EMF (2015). While the absence of a hierarchy makes it easier to adapt the recommended principles and strategies to different situations, it has the counterpart of putting on the same level the benefits that can be expected from waste recyling on the one hand and from the reduction of materials and waste flows on the other hand.

However, research shows that even if the recycling rates of construction and demolition waste increase, secondary resources can only partially replace primary resources. In Vienna, if all construction and demolition waste was treated in a recycling center, the consumption of primary resources would only be reduced by 7% (Obernosterer *et al.* 1998). In Orléans (France), recycling 70% of mineral waste would only cover a quarter of aggregates consumption from 2005 to 2030 (Serrand *et al.* 2013). In 27 European Union countries, this recycling rate would only meet half of the required consumption by 2020 (Wiedenhofer *et al.* 2015).

Therefore, a circular economy model for construction cannot be achieved solely by a change, however strong it may be, in the practices of recycling construction and demolition waste. An absolute reduction in inflows (as well as outflows), as advocated by the transformationist school of the circular economy identified by Reike *et al.* (2018), is necessary to address the challenges associated with the depletion of materials resources. It should also be noted, as Gemeente Amsterdam *et al.* (2019) point out, that energy resources also present strong constraints and limit, in particular, the possibilities of recycling waste.

However, extending the limits of the scope of the circular economy to confuse its limits with those of sustainable development, as proposed by Pomponi and Moncaster (2017), presents the risk of removing the operational characteristic of the circular economy, which is one of the main strengths of this concept (Geissdoerfer *et al.* 2017). As such, a balance must be found between an operational concept aiming at responding to the material and energy challenges facing human societies by reintegrating the activities of these societies within the limits of the biosphere and sustainable development, which has broader purposes.

10.5.2. *Low consideration of the territorial context*

The second limitation that can be identified is the universality of many of the definitions of the circular economy for construction and of the tools for its application. Indeed, very few definitions refer to local construction issues, and in particular to issues related to the availability of resources for construction. In addition, some tools aiming at taking into account the local context are largely based on data from national average values, as well as in CSTB *et al.* (2018) and Oury (2019).

However, the availability of natural resources (known as primary resources) on the one hand and anthropogenic resources (known as secondary resources) consisting of waste resulting from the renewal or demolition of built works on the other hand varies greatly according to the territories. The same is true of the potential for the substitution of primary resources by secondary resources. Brunner (2011) considers in particular that three phases of urban development must be distinguished: availability and potential are lower for fast-growing cities; they are higher for shrinking cities experiencing demographic decline, while mature urbanized areas undergoing high renewal are in an intermediate situation.

Among the documents studied, the recommendations for the Brussels region by Athanassiadis (2017) are an exception as they relate the current situation of the region to the desired situation. These recommendations are based on the analysis of flows and stocks of materials and energy, an observation also known as the study of metabolism and using methods developed in the scientific fields of industrial and territorial ecology. The operational concept of the circular economy could be better linked to the analytical concept of the metabolism to define strategies that take better account of the territorial context.

10.5.3. *A scale for the application of the circular economy concept that is too narrow*

The third limitation that can be observed is the scale defined to apply the concept of the circular economy. A majority of the definitions studied, as well as a large proportion of policies and projects, are limited to the scale of the construction site. Targeting new construction may be relevant in the case of fast-growing cities. However, in already urbanized areas, the buildings resulting from new construction represent relatively small surfaces compared to the existing buildings. As noted by Pomponi and Moncaster (2017), more than three-quarters of the built stocks currently present in developed countries will still be present in 2050. In these territories, the scale of the individual building must be exceeded in order to transform the entire existing built area.

In addition, the barriers to the implementation of circular economy strategies for construction take place at multiple scales, from the construction site or project to the city, region, country, European Union or even a larger international scale. This is particularly the case for the barriers to the use of secondary resources in construction in France, which fall within the insurance and regulatory framework (Augiseau 2020). Acting only at the scale of the construction site will not remove these constraints. As ARUP (2016) observes, a circular economy can only be achieved if strategies integrate different scales in a coherent way.

10.5.4. *Insufficient coordination between the circular economy and spatial planning*

The fourth limitation that can be observed is the lack of spatialization of the strategies implemented and their weak coordination with spatial planning. This low spatialization is related to the two previous limitations. It also results from the transposition of ideas originally formulated to improve the production of manufactured goods (Obersteg *et al.* 2019). The lack of coordination between circular economic strategies led by cities and spatial planning is observed by Petit-Boix and Leipold (2018). Obersteg *et al.* (2019) make the same observation for the six regional strategies in Europe they analyzed.

While an "urban plan or project also has a volume [...], mass and energy content" (Barles 2015, author's translation), the impact of urban plans and projects on materials consumption and waste generation is rarely a decision-making criterion for urban actors (Kennedy *et al.* 2011). However, the work led by the International Resource Panel of the United Nations Environment Programme (IRP 2018) recommends reducing materials flows by influencing urbanization processes. This work highlights in particular the risk of high urban sprawl in developing countries.

Reducing materials flows "requires rethinking the shape of urban agglomerations to minimize infrastructure stocks [and] reducing resource consumption induced by the structure and spread of the urban fabric" (Erkman 2004, p. 182, author's translation). Urban planning can also ensure that the waste flows resulting from demolition are limited by giving priority to the refurbishment of buildings.

Local authorities can also promote the use of secondary resources by linking urban planning with stocks and materials flows planning. The creation of a spatialized database or cadastre of secondary resources (Brunner 2011) can provide a basis for coordinating urban projects in order to promote flow exchanges between sites. The Plaine Commune Urban Metabolism project (Bellastock *et al.* 2018), as well as the Est Ensemble project (Augiseau 2020), aim to build such bases. Urban planning and urban development are means of action on urban metabolism and can lead to reducing materials flows. Circular economy policies and projects could be strengthened by integrating spatial planning into their scope of action.

10.6. Conclusion

The literature review on the concept of the circular economy applied to construction identified sixteen elements of definition. A great heterogeneity can be observed among these elements, but the recommendations to anticipate and apply the recycling of construction and demolition waste at the building scale dominate. While this heterogeneity seems to reflect a lack of stability in the concept of the circular economy for construction, a proliferation of policies, research and development projects as well as construction and urban development projects is nevertheless observed. The circular economy strategies that were reviewed could be a response to the tough challenges that the construction industry faces, but they should overcome four limitations observed in the definitions, policies and projects: the uncertain delimitation of the scope of the circular economy, the low consideration of the territorial context when defining strategies, the too narrow scale for the application of the circular economy concept, and the lack of coordination between the circular economy and spatial planning.

10.7. References

Adams, K., Osmani, M., Thorpe, T. and Thornback, J. (eds) (2017). Circular economy in construction: Current awareness, challenges and enablers. In *Proceedings of the Institution of Civil Engineers: Waste and Resource Management*, ICE Publishing, pp. 15–24.

ADEME (2014). Economie circulaire : notions. Technical file, ADEME.

ADEME (2018). Economie circulaire dans le BTP. Bilan des 18 projets de R&D de l'APR, ADEME, 2012–2014.

ARUP (2016). The Circular Economy in the Built Environment. Presentation, ARUP.

Athanassiadis, A. (2017). Economie circulaire dans le secteur de la construction à Bruxelles : état des lieux, enjeux et modèle à venir. Available at: https://www.researchgate.net/publication/321156948_Economie_Circulaire_dans_le_secteur_de_la_Construction_a_Bruxelles_Etats_des_lieux_enjeux_et_modele_a_venir.

Augiseau, V. (2017). La dimension matérielle de l'urbanisation. Flux et stocks de matériaux de construction en Ile-de-France. PhD thesis, Université Panthéon-Sorbonne-Paris I.

Augiseau, V. (2020). Utiliser les ressources secondaires de matériaux de construction : contraintes et pistes d'action pour des politiques territoriales. *Flux*, 116–117, pp. 21–36.

BAM, BRE, cd2e, London Waste & Recycling Board, Ouroboros, Tarkett and Turntoo (2016). Circular economy in the built environment: Case studies. Available at: https://www.ellenmacarthurfoundation.org/assets/downloads/Built-Env-Co.Project.pdf.

Barles, S. (2015). L'urbanisme, le génie urbain et l'environnement : une lecture par la technique. *Riurba*, 5(1).

Bellastock, Albert & Compagnie, Auxilia, Recovering, Le Phares, CSTB, Encore Heureux, Halage and BTP Consultants (2018). Bilan technique et stratégique. An 1 de la phase opérationnelle du projet Métabolisme urbain. 2017–2018.

Brand, S. (1994). *How Buildings Learn: What Happens After They Are Built*. Viking Penguin, New York.

Bringezu, S. (2002). Towards sustainable resource management in the European Union. Wuppertal Institute for Climate, Environment and Energy, Germany.

Brunner, P.H. (2011). Urban mining, a contribution to reindustrializing the city. *Journal of Industrial Ecology*, 15(3), pp. 339–341.

Circle Economy, DGBC, Metabolic, SGS Search and REDEVCO Foundation (2018). A Framework for Circular Buildings: Indicators for Possible Inclusion in BREEAM. Circle Economy. Available at: https://www.circle-economy.com/insights/a-framework-for-circular-buildings.

CSTB, BRGM, Bouygues Construction and Alliance HQE-GBC (2018). Manuel de l'utilisateur. Outils pour VALoriser les actions de transition vers une Economie Circulaire dans la construction – focus sur les flux matériaux / déchets de bâtiment. Available at: http://www.hqegbc.org/wp-content/uploads/2017/05/190118_livret-utilisateur-OVALEC.pdf.

CSTC (2017). Construire circulaire – vers une économie circulaire dans la construction. Innovation Paper, CSTC.

van den Dobbelsteen, A. (2008). 655: Towards closed cycles – New strategy steps inspired by the Cradle to Cradle approach. *25th Conference on Passive and Low Energy Architecture*, Dublin.

Duffy, F. (1990). Measuring building performance. *Facilities*, 8(5), pp. 17–20.

Elmqvist, T., Fragkias, M., Goodness, J., Güneralp, B., Marcotullio, P.J., McDonald, R.I., Parnell, S., Schewenius, M., Sendstad, M., Seto, K.C., Wilkinson, C. (eds) (2013). *Urbanization, Biodiversity and Ecosystem Services: Challenges and Opportunities: A Global Assessment.* Springer, Berlin.

EMF (2015). Growth within: A circular economy vision for a competitive Europe. Available at: https://www.ellenmacarthurfoundation.org/assets/downloads/publications/EllenMacArthurFoundation_Growth-Within_July15.pdf.

EMF (2017). Cities in the circular economy: An initial exploration. Available at: https://www.ellenmacarthurfoundation.org/assets/downloads/publications/Cities-in-the-CE_An-Initial-Exploration.pdf.

Erkman, S. (2004). *Vers une écologie industrielle. Comment mettre en pratique le développement durable dans une société hyper-industrielle*, 2nd edition. Edition Charles Léopold Mayer, Paris.

European Commission (2019). Report from the Commission to the European Parliament, the Council, the European Economic and Social Committee and Committee of the Regions on the implementation of the Circular Economy Action Plan.

European Commission Joint Research Centre (2017). Level(s) – A common EU framework of core sustainability indicators for office and residential buildings. Parts 1 and 2: Introduction to Level(s) and how it works (Draft Beta v1. 0).

European Economic and Social Committee (2019). Circular economy strategies and roadmaps in Europe: Identifying synergies and the potential for cooperation and alliance building. Final Report.

Fondation Bâtiment Energie (2018). Appel à manifestation d'intérêt sur l'Economie Circulaire. Elaboration de critères et indicateurs pour le développement de bases scientifiques à la caractérisation de l'économie circulaire dans le secteur du bâtiment.

Geissdoerfer, M., Savaget, P., Bocken, N.M.P. and Hultink, E.J. (2017). The Circular Economy – A new sustainability paradigm? *Journal of Cleaner Production*, 143, pp. 757–768.

Geldermans, R.J. (2016). Design for change and circularity – Accommodating circular material & product flows in construction. *Energy Procedia*, 96, pp. 301–311.

Gemeente Amsterdam, SGS Search and Metabolic (2019). Roadmap circular land tendering. An introduction to circular building projects.

Geraedts, R.P. and Prins, M. (2015). The CE Meter: An instrument to assess the circular economy capacity of buildings. *CIB International Conference Going North for Sustainability*, London.

Ghisellini, P., Cialani, C. and Ulgiati, S. (2016). A review on circular economy: The expected transition to a balanced interplay of environmental and economic systems. *Journal of Cleaner Production*, 114, pp. 11–32.

Gladek, E. (2017). The seven pillars of the circular economy. Available at: http://www.metabolic.nl/the-seven-pillars-of-the-circular-economy (accessed November 2, 2019).

Gordon, R., Bertram, M., and Graedel, T.E. (2006). Metal stocks and sustainability. *Proceedings of the National Academy of Sciences*, 103(5), pp. 1209–1214.

Hashimoto, S., Tanikawa, H. and Morigushi, Y. (2007). Where will large amounts of material accumulated within the economy go? A material flow analysis of construction minerals for Japan. *Waste Management*, 27, pp. 1725–1738.

Hobbs, G. (2019). Circular Building Assessment – A new tool made possible by BAMB. Report. BAMB, Brussels.

IRP (2018). The weight of cities: Resource requirements of future urbanization. Report, International Resource Panel. United Nations Environment Programme, Nairobi, Kenya.

Kennedy, C., Pincetl, S. and Bunje, P. (2011). The study of urban metabolism and its applications to urban planning and design. *Environmental Pollution*, 159(8), pp. 1965–1973.

Krausmann, F., Gingrich, S., Eisenmenger, N., Erb, K.-H., Haberl, H., and Fischer-Kowalski, M. (2009). Growth in global materials use, GDP and population during the 20th Century. *Ecological Economics*, 68(10), pp. 2696–2705.

Krausmann, F., Wiedenhofer, D., Lauk, C., Haas, W., Tanikawa, H., Fishman, T., Miatto, A., Schandl, H. and Haberl, H. (2017). Global socioeconomic material stocks rise 23-fold over the 20th Century and require half of annual resource use. *Proceedings of the National Academy of Sciences*, 114(8), pp. 1880–1885.

Leising, E., Quist, J. and Bocken, N. (2018). Circular Economy in the building sector: Three cases and a collaboration tool. *Journal of Cleaner Production*, 176, pp. 976–989.

Mairie de Paris (2017). Plan économie circulaire de Paris 2017–2020. 1ère feuille de route.

McDonough, W. and Braungart, M. (2002). *Cradle to Cradle – Remaking the Way We Make Things*. North Point Press, New York.

Ministère de la transition écologique et solidaire, Ministère de la cohésion des territoires, ADEME, CSTB and Association OREE (2017). Cadre de définition de l'économie circulaire dans le bâtiment.

Ministère de la transition écologique et solidaire and Ministère de l'économie et des finances. (2018). Feuille de route économie circulaire. 50 mesures pour une économie 100% circulaire.

Mishra, S. and Siddiqui, N.A. (2014). A review on environmental and health impacts of cement manufacturing emissions. *International Journal of Geology, Agriculture and Environmental Sciences*, 2(3), pp. 26–31.

Obernosterer, R., Brunner, P.H., Daxbeck, H., Baccini, P., Deistler, M., Lahner, T., Lohm, U. and van der Voet, E. (1998). Materials accounting as a tool for decision making in environmental policy: MAcTEmPo case study report. Urban metabolism, Vienna.

Obersteg, A., Arlati, A., Acke, A., Berruto, G., Czapiewski, K., Dąbrowski, M., Heurkens, E., Mezei, C., Federica Palestino, M., Varjú, V., Wójcik, M. and Knieling, J. (2019). Urban regions shifting to circular economy: Understanding challenges for new ways of governance. *Urban Planning*, 4(3), pp. 1–13.

OECD (2018). Global Material Resources Outlook to 2060 – Economic Drivers and Environmental Consequences. Highlights.

O'Rourke, D., Connelly, L. and Koshland, C. (1996). Industrial ecology: A critical review. *International Journal of Environment and Pollution*, 6 (2/3), pp. 89–112.

Oury, N. (2019). Réunion de lancement. Test HQE Performance Economie circulaire. Presentation, 12 March 2019, Paris.

Peduzzi, P. (2014). Sand, rarer than one thinks. *Environmental Development*, 2014/11, pp. 208–218.

Petit-Boix, A. and Leipold, S. (2018). Circular economy in cities: Reviewing how environmental research aligns with local practices. *Journal of Cleaner Production*, 195, pp. 1270–1281.

Pomponi, F. and Moncaster, A. (2017). Circular economy for the built environment: A research framework. *Journal of Cleaner Production*, 143, pp. 710–718.

Pullen, S.F. (2000). Energy used in the construction and operation of houses. *Architectural Science Review*, 43(2), pp. 87–94.

Reike, D., Vermeulen, W.J.V. and Witjes, S. (2018). The circular economy: New or refurbished as CE 3.0? Exploring controversies in the conceptualization of the circular economy through a focus on history and resource value retention options. *Resources, Conservation and Recycling*, 135, pp. 246–264.

Serrand, M., Monfort-Climent, D., Vaxelaire, S., Michel, P. and Gautier, A.-L. (2013). Projet ANR ASURET – Méthodologie d'aide à la décision : evolution prospective des besoins en matériaux et de la disponibilité des ressources primaires et secondaires – Exemple de la Ville d'Orléans (tâche 5). Final Report BRGM/RP-62586-FR.

Sommer, J. and Guldager, K. (2016). *Building a Circular Future*. GXN Innovation, Copenhagen

UKGBC (2019). Circular economy guidance for construction clients: How to practically apply circular economy principles at the project brief stage. Report, The Crown Estate.

Vincent, M. (1985). Les secteurs de la construction, un panorama des analyses récentes et des voies de recherches. *Revue d'économie industrielle*, 32 (2nd quarter), pp. 86–109.

Wiedenhofer, D., Steinberger, J.K, Eisenmenger, N. and Haas, W. (2015). Maintenance and expansion: Modeling material stocks and flows for residential buildings and transportation networks in the EU25. *Journal of Industrial Ecology*, 19(4), pp. 538–551.

11

Understanding the Concept of Waste to Avoid its Production

11.1. Introduction

Even if the rhetoric of the circular economy is based on the reference to the "new", let us not forget the genesis of "zero waste" in light of the mythical "zeros" of the "Japanese moment" and the ideology of Toyotism, with reference to the five "zeros": "Zero defect", "Zero paper", "Zero weakening", "Zero stock" and "Zero delay" (De Rozario and Pesqueux, 2018), an ideology of perfection in some way.

If the aim of the circular economy is to achieve a waste-free socio-economic functioning, then it is essential to ask the question of an ontology of waste, without which it would not be possible to establish the ontology of a circular economy, except in a formal way.

The ontology discussed in this chapter assumes that waste is what crosses a boundary, thus making it possible to characterize the area from which it comes, as well as the area where it goes, both being in duality.

It is possible to consider waste as one of the analyzers of how a society functions insofar as it is representative of social relations, class differences, gender duality (male waste can be substantially different from female waste), cultural differences between societies (see the archaeology of Paleolithic waste, which allows us to try to understand them in the absence of any written trace) or conceptions of hygiene that form the basis for the difference between clean and dirty; the dirty being the first step in a kind of waste lifecycle, this lifecycle being the descriptor used to define waste (its epistemology).

Chapter written by Yvon PESQUEUX.

From an etymological point of view, the French word for waste, "*déchet*", derived from the verb *déchoir*, meaning decay, meant, from the Middle Ages onwards, the part that was lost. At the beginning of the 19th Century, the word also characterized a degenerate person. Anecdotally, let us note how the Japanese housewife describes her spouse reaching retirement age as *dai gomi* (big waste) – a sign of his loss of utility and social status and fear of seeing him too present at home!

In the tradition of Serres (1990), let us recall the book by Montsaingeon (2017) which notes that waste is an indelible sign of our presence on Earth as much as the symptoms of the current world crisis, hence the injunction to reduce, reuse and recycle it. At a time when reference is being made to the circular economy, this utopia of a non-decaying mode is reminiscent of the Chagga tribe's lie, evoked by Douglas (1966): adult men of this tribe claim never to defecate. Montsaingeon's book highlights the quest for purity, built on the utopia of a technical mastery of waste. So, utopia or dystopia? It tells how *Homo detritus* constitutes the dual face of *Homo economicus* building the belief in a world that could be saved by throwing away properly insofar as the circular economy would replace NIMBY (not in my backyard).

And yet, when the human being is considered as a resource (through their skills and labor force), they can also end up as waste. This is traditionally true for certain professions and sectors of the economy, as well as for age, health status, etc. With regard to the notion of "human capital", it is a question of return on investment, depreciation and therefore their passage to a relegation zone (less important duties, marginal roles, etc.) or directly to waste status (early retirement, etc.). For example, there is talk of "refreshing the age pyramid" in multinational companies. The organizational agent can therefore be perceived as a "potential waste" with regard to two references: their productivity and their "recycling" potential. This externality is also managed differently according to the company (for example, with pre-retirement status in France or the return of dependent children).

The relationship with waste can be a challenge for institutional construction (1975 in France with the emergence of the Ministry of the Environment). It raised the question of the relationship between biological society (the universe of the family and its domestic waste), civil society (in its economic dimension with industrial waste, non-economic with the debate on the place of NGOs in the theme of waste) and political society (existence of a Ministry in charge of the issue, public policies expressed in terms of subsidies, incentives and the alternative between collection and what happens afterwards, in terms of environmental protection).

As a descriptor, waste is representative of the inequalities at work within a society. In line with the "fuzzy and ambiguous" dimension of sustainable development (Pesqueux 2011), we find, with waste, this double aspect: why and when does a good become waste? The ontology of waste would then be metonymic with that of society, where waste is considered as an essential residue of its functioning, and also as an "object" that gives rise to the construction of a political discourse (what is the attitude towards the waste that is most often to be discarded, given the whole issue of the efficiency of disposal) and moral (waste is not good and it is in this sense that it is called garbage, a concept with negative connotations). The reference to waste is a prism which allows us to enter into the dual dimension of balance and harmony in a society.

The text is therefore based on two arguments: waste is what crosses a boundary (its ontology) and waste is defined by a lifecycle (its epistemology), the passage from its ontology to its epistemology being carried out in light of its ambiguity, whose fresco of institutional definitions tends to build a Prévert inventory[1].

11.2. Waste defined as that which crosses a boundary

How can we define what waste is? When does an object become waste? While a product is an "object" for some, isn't it already "waste" for others? Through the literature, waste does not seem to be the subject of an accepted definition, except for institutional purposes. It is possible to hypothesize that our perception of and relationship to waste depends on the social, economic and cultural context. In addition, waste is linked to the means put in place at a given place and time to get rid of it.

In mass consumer societies, waste is generally perceived as normal production. It is in this context that the logic of waste recovery has developed with regard to the notion of sustainable development, even if the link is not as obvious as it seems. In developing countries, waste is often considered as a resource and a means of obtaining an income for those who collect it. "Everything has a value, a use and man still controls the cycle of materials" (Chalmin and Gaillochet, 2009). The European Union proposes the following definition: "any substance or object which the holder

1 Jacques Prévert was a 20th-Century French poet. He was famous for accumulating words at random in his poems, and hence the expression "a Prévert inventory" is used, in which we may find very different things gathered at random.

discards or intends or is required to discard"[2], and in which the criterion of "uselessness" or "utility" is decisive (*a first boundary*).

It is this crossing that now forms the basis for the transition from a status of free private property to a public good that can be valued (the world upside down!) (*a second boundary*), which forms the basis of the trilogy "recover–recycle–valorize". Once considered by its owner as useless, the good that has become private waste becomes, by rejection, a public good. It is then most often placed within the public domain and remains there until the *ad hoc* services or time and the elements take it to another destination. In the context of individual property, "public space functions as a periphery of the habitat, receives the waste and causes it to fade away, under the influence of rain or traffic" (Botta *et al.*, 2002).

First, the notion of waste, to be valid, is based on the existence of an interior and an exterior (*a third boundary*), waste being what crosses the boundary from the interior to the exterior as if it were a return to nature. By crossing this boundary, it ceases to be the property of the person who rejects it.

This process of passage feeds the "willing-guilt" duality (*a fourth boundary*), unwillingness when it comes to rejecting without any other form of trial, willingness when it comes to sorting, to conditioning, although it is a rejection. This notion is implicitly proprietarist with regard to the *usus* – which has led to its term, the *fructus* – which has been removed from use and abuse – materialized by the rejection. Ownership is the way this boundary is constructed, waste being what crosses the private boundary (as a result of a structured approach) to go to the public stage (where it becomes unstructured). It is the passage between these two universes that forms the basis of its genesis it and induces the waste lifecycle.

For Bertolini (1990), waste is "nomadic", its status as waste being only temporary (*a fifth boundary*), the status of waste excluding it from the universe of sustainable to send it into that of unsustainable. For example, with recycling, it evolves, changes in nature and status: a plastic bottle is reincarnated into a polar garment, a tetra pack of milk as a handbag or organic waste as fertilizer in the garden. The recovery process is built on a transition from the temporary state of waste, to a sustainable state of resource. Recycling is at the "boundary of which waste is no longer" (Chalmin and Gaillochet, 2009).

2 Article 3.1 Directive 2008/98 EC of the European Parliament and of the Council of November 2008 on waste.

11.3. The ambiguity of waste

Far from the eyes of the beholder, its future is of little concern to them. It is only in light of local legislative progress and normative work, of the type initiated by the OECD on extended producer responsibility (EPR), that it is a question of restoring private responsibility (to the waste producer) by extending the "polluter-pays" type of legislation that has developed for industrial and agricultural activities. We could even describe this EPR as a sign of tension between defeatism (we have to "do with" waste) and struggle. Conversely, when it is recovered in landfill by a reclaimer or through a formal recovery cycle, can the recycled waste therefore return to the private sphere (a privatization), a kind of archetype of the circular economy?

With regard to the previous argument, it should be noted that there is a duality in the value judgment for waste: in the duality of those of EPR, there are those in the shadows with degassing, the release of hazardous materials. This duality is also a sign of the overdose of a contaminated society.

Hence, waste for some people is not necessarily waste for others. Waste therefore depends on social class, and in an omission, we could say that class waste is also a founding act of waste class(ification). We then understand very well that the "picking up – collection" of waste from the rich is more valuable than that of the poor ("tell me what you throw away, I will tell you who you are"). But is the economic value enough to support a class analysis on this subject?

Waste involves "picking up – collection", i.e. a process of the type, "collection by loading – discharge by dumping". It is this process that now underlies the idea of the transition from collection (the process of taking waste and which will be fundamentally linked to its nature) to collection, this second name containing seeds of the idea of recovery. It is with the collection process that we talk about technology that links garbage cans, monstrous, bulky objects, "green" or vegetable waste, toxic waste and the method of collection, landfill and incineration.

Waste is an object considered dirty and should be disposed of and removed from the clean private space and thus purified (Botta *et al.* 2002). It is in response to this that companies and public authorities have realized its usefulness and potential. They summarize this ambiguity as follows:

> "We will need to take waste into account in a new way, and take a greater interest in it to resolve this contradiction, the tension that exists between the desire to reject this troublemaker and the realization that the blind rejection of garbage is a serious risk factor for our

> environment and our development; whose sustainability is no longer obvious" (pp. 11–12).

Waste is not only the excreta that is essential and is treated as such. The notion is also valid with that of scum, the other form of waste marked by the idea that we no longer want it, that we must get rid of it. It is on this basis that the themes of recovery, reuse and recycling attempt to build a return from the outside to the inside. There is much less question, in these logics, of reducing them.

It is at the heart of this process that economic redemption lies with the notion of "waste value", which will depend on several factors: the scarcity of the raw material from which the waste is derived, the profit that the people involved in the chain can make (Wilson *et al.* 2006, p. 801), the standard of living of the individuals involved in the chain (here we come back to the idea that waste can be a resource for some, waste for others), the costs associated with its recovery or disposal and the level of sorting (the more waste is sorted, the more economically valuable it is) (Hoornweg and Bada-Tata 2009, p. 801).

Chalmain and Gaillochet (2009) distinguish two types of waste: waste whose exchange value is negative and does not represent any economic interest, particularly if its recovery involves a cost greater than the cost of disposal, a negative externality then, and waste whose use and exchange value is positive and which can be transformed into matter or energy according to its recoverable content.

The valuation process is used to create added value for a product that has no original value. At each stage of the value chain, waste becomes a resource whose price is increasingly negotiable.

11.4. Institutional definitions of waste

The fresco of definitions, functionalist in nature, lies at the limit of a Prévert inventory between:

– *waste defined by the Basel Convention*[3] on the Control of Transboundary Movements of Hazardous Wastes and Their Disposal, is "substances or objects that are disposed of, intended to be disposed of or are required to be disposed of under the provisions of national law";

---

3 Basel Convention, Article 2, paragraph 1, https://www.basel.int/Portals/4/Basel%20Convention/docs/text/BaselConventionText-e.pdf. Adopted by the Plenipotentiary Conference on March 22, 1989, it entered into force on May 5, 1992.

– *solid waste is defined by the United Nations Agenda 21*[4] as "all domestic refuse and non-hazardous waste such as commercial and institutional waste, street sweepings and construction debris";

– *the definition of the French Environmental Code* (Title IV of Book V – Prevention of pollution, risks and nuisances), which defines waste as "any substance or object, or more generally any item, of which the holder is disposing or of which he has the intention or the obligation to dispose" (Article L 541-1-1)[5];

– waste recovery is defined by Directive 2008/98/EC of the European Parliament and of the European Council of November 19, 2008, on waste as "any operation the principle result of which is waste serving a useful purpose by replacing other materials which would otherwise have been used to fulfil a particular function, or waste being prepared to fulfil that function, in a factory or in the wider economy". More specifically, the French organization ADEME (*Agence de l'environnement et de la maîtrise de l'énergie*) presents waste recovery as "a generic term covering the reuse, repurposing, regeneration, recycling, organic recovery or energy recovery of waste"[6];

– *household waste* concerns all individuals, in their different aspects as citizens, inhabitants and consumers; the French Environmental Code defines it, in article R 541-8, as "any waste, hazardous or non-hazardous, the producer of which is a household"[7].

Within household waste, we distinguish between:

– *household waste*: waste resulting from the domestic activity of households and which can be sorted (selective collection) or not (we refer here to residual household waste);

---

4 Agenda 21, Section 21.3, https://sustainabledevelopment.un.org/content/documents/Agenda21.pdf.

5 Environmental Code resulting from Ordinance No. 2000-914 of September 18, 2000, on the legislative section of the Environmental Code. Books I, III, IV and V of the regulatory section of the Environmental Code were published by Decree No. 2005-935 of August 2, 2005, on the regulatory section of the Environmental Code and Books II and VI were published by Decree No. 2007-397 of March 22, 2007.

6 www2.ademe.fr.

7 INSEE, Definition, method and quality, "In general, a household, in the statistical sense of the term, refers to all the occupants of the same dwelling without these persons necessarily being related to each other (in the case of cohabitation, for example). A household can be composed of only one person": http://www.insee.fr/fr/methodes/default.asp?page=definitions/menage.htm, accessed on 15/07/19.

– *bulky objects*: domestic waste which, because of its volume or weight, cannot be taken into account by the usual collection of household waste and requires a specific management method;

– *special waste*: this is domestic hazardous waste (paints, solvents, batteries, etc.) that requires a specific management method[8];

– "household waste" (HW) is relatively similar to and often associated with "household and similar waste" (HSW), which includes assimilated waste (or garbage).

The functional typology of waste is as follows:

– *municipal solid waste*, which includes household and similar waste, special household waste, waste from roads, markets and public places, green waste from public spaces, agricultural waste, wood waste and ordinary industrial waste. They can be classified as putrescible waste (food waste, leftovers, garden waste), paper, cardboard, glass, plastics, textiles, wood and metals, scrap metal, copper pieces, aluminum scrap, etc.;

– *special household waste* is mainly composed of batteries, light bulbs, solvents, paint, varnish, glues, batteries, toner cartridges for printers, fluorescent tubes, cleaning products, aerosols and plant protection products;

– *toxic waste* in dispersed quantity is the same as the above waste but is held by industries. Products containing organic and metallic micro-pollutants are then added to the previous list. With special household waste, they have a potential risk because they are heterogeneous, sources of disparate hazards and are subject to specific selective sorting policies, specific collection and treatment processes;

– *non-hazardous waste* from economic activities comes from industrial production and is assimilated, to the nearest scale, to household waste;

– *electrical and electronic equipment waste* (or e-waste for the electronic part).

The issue of waste is part of different levels of international thinking. It has been incorporated into the Sustainable Development Goals (SDGs) adopted at the Summit on Sustainable Development (September 25, 2015) by UN Member States[9] with regard to three objectives: to halve the amount of food waste per capita globally by 2030; to achieve environmentally sound management of chemicals and all waste

8 ADEME, Report "*Déchets - Edition 2015*".
http://www.ademe.fr/sites/default/files/assets/documents/chiffres-cles- waste-201507_8500.pdf.

9 United Nations Development Program, SDG, Goal 12, "Ensure sustainable consumption and production patterns".

throughout their lifecycle and to significantly reduce waste generation through prevention, reduction, recycling and reuse by 2030.

At the European level, the development and implementation of waste management legislation and policies is carried out within the framework of several European policies and programs such as the EU's "7th Environment Action Programme (EAP)"[10], the "Europe 2020 Strategy" or the "Roadmap to a Resource Efficient Europe"[11]. In December 2015, a new legislative package "Circular Economy"[12] was proposed and presented to the European Parliament in June 2016. The Waste Framework Directive 2008/98/EC is the subject of most of the amendments tabled, as this Directive is directly linked to all the others (Directive 94/62/EC on packaging and packaging waste, Directive 1999/31/EC on landfill waste, Directive 2000/53/EC on end-of-life vehicles, Directive 2006/66/EC on batteries and accumulators and Directive 2012/19/EC on waste electrical and electronic equipment).

In France, waste management is also part of several programs and levels of obligation. The *Programme national de prévention des déchets 2014–2020* (French National Waste Prevention Program 2014–2020) aims to gradually break the link between economic growth and waste production, by extending previous prevention actions and setting new targets such as a 7% reduction in DMAs compared to 2010 and the stabilization of business waste. In addition to this program, the *Plan de reduction et de valorisation des déchets 2014–2020* (French Waste Reduction and Recovery Plan 2014–2020)[13] reinforces the role of waste policy in the transition to a circular economy with targets for 2020: a 10% reduction in DMA produced per capita (compared to 2010), the recovery of non-inert hazardous waste materials by 55% in 2020 and 60% in 2025 and a 30% reduction in landfill tonnage by 2020 and 50% by 2025. The *Loi relative à la Transition Energétique pour la Croissance Verte* (French Law on Energy Transition for Green Growth) (Law No. 2015-992 of August 17, 2015[14]), in its Title IV, "Combating waste and promoting the circular economy: From product design to recycling", sets

10 European Commission, eea.europa.eu/policy-documents/7th-environmental-action-programme.

11 European Commission, https://ec.europa.eu/environment/resource_efficiency/about/roadmap/index_en.htm.

12 European Commission, https://ec.europa.eu/environment/circular-economy.

13 French Ministry of Environment, Energy and Sea, "*Plan de réduction et de valorisation des déchets 2014–2020*", http://www.developpement-durable.gouv.fr/IMG/pdf/14201_RV-plan-dechet-gd-public_BATlight-2.pdf.

14 Legifrance, https://www.legifrance.gouv.fr/affichTexte.do?cidTexte=JORFTEXT000031044385&categorieLien=id5.

the objective of going beyond the linear economic model and affirms the essential role of national waste prevention and management policy in achieving this.

The organization of waste management is based on the principle of extended producer responsibility (EPR). Promoted by the Organisation for Economic Co-operation and Development (OECD) in the 1990s, the producer of a product is responsible for the product throughout its lifecycle. One of the objectives is to internalize environmental costs in the price of new products, including collection, recycling and treatment costs. It should also encourage the producer to take environmental aspects into account from the product design stage, in order to prevent waste generation at source and to facilitate recycling[15]. There are currently 14 mandatory EPR programs. Seven of them are imposed by a European directive or in response to a European directive or a Community regulation (batteries and accumulators, electrical and electronic equipment – DEEE, cars, lubricants, household packaging, fluorinated refrigerants and medicines). The other seven are imposed by national regulations (tires, graphic papers, textiles/homeware and footwear, furniture, chemicals, infectious risk care activities and gas bottles). In addition to the EPR programs imposed by regulations, voluntary agreements have been set up between economic actors and public authorities (plastic packaging for agricultural supplies, printing cartridges and mobile homes). Producers are responsible for the collection and treatment of their products at the end of their lifecycle, and, faced with diffuse and composite volumes, they have joined forces to set up organizations – eco-organizations – to which they delegate these responsibilities and which occupy a central place in the EPR programs with a dual mission: a political mission of general interest for the collection and treatment of waste in compliance with national and European regulations, and an economic and managerial mission focused on seeking efficiency, reducing costs and optimizing resources. There are two types of eco-organizations: eco-organizations whose actions consist of collecting the eco-contributions due by the marketers and of paying financial support to certain actors, such as local authorities (this is the case for packaging and graphic papers), and so-called operational eco-organizations if the producer's responsibility they assume concerns the collection and treatment of used products. In this case, they use service providers selected by invitation to tender (e.g. tires, batteries and accumulators or DEEE[16]).

15 *Ministère de l'Environnement, de l'Energie et de la Mer*, "*Principe de Responsabilité élargie du producteur*", http://www.developpement-durable.gouv.fr/Le-principe-de-la-responsabilite,12046.html.

16 Cottel and Chevrollier, (2013) "*La gestion des déchets dans le cadre des filières à responsabilité élargie des producteurs*", *Assemblée Nationale*, http://www.assemblee-nationale.fr/14/rap-info/i1347.asp.

11.5. Lifecycle analysis

This is based on a phased decomposition:

– *pre-collection*, which is all the operations organizing the disposal of waste from the place of its production to its acceptance by an approved municipal or private service;

– *recovery* that is carried out between pre-collection and collection and corresponds to all the operations organized for recovery. It is carried out through two channels: direct collection from households and industry or from collection points, recovery from landfill, etc.;

– *collection*, which includes all organized waste collection action by any natural or legal person authorized for this purpose. The European directive[17] defines collection as "the gathering of waste, including the preliminary sorting and preliminary storage of waste for the purposes of transport to a waste treatment facility". The idea of sorting is present here. Collection activity depends on the topography of the areas covered, population density, the nature of the issuers (households, companies due to quantities) and the condition of the infrastructure. It is based on two main rationales: fixed-point collection (agents voluntarily deposit waste there, collection takes place at regular intervals to take its contents to sorting and/or treatment centers) and door-to-door collection, which involves the direct transfer of waste inside collection vehicles;

– *reuse*, which is defined as the use of waste by another entity and/or for a different use (plastic bags which then serve as a garbage bag). The flea market is a method of organizing reuse. It consists of re-injecting objects that have already been used into the market at prices much lower than those of equivalent materials. This is also referred to as "second hand". The flea market offers a double advantage: an additional income for the entities that sell them and an affordable price for those that buy them;

– *recycling*, which is the "recovery operation by which waste materials are reprocessed into products, materials or substances whether for the original or other purposes"[18]. Some materials are more easily recyclable than others. Recycling is generally the result of a "refining" process leading to a homogeneous raw material, refining also inducing waste and pollution. It is also about upcycling, which is a reuse to make something else;

---

17 European Directive, https://eur-lex.europa.eu/legal-content/EN/TXT/?uri=CELEX:32008L0098.

18 European Directive, *ibid.*

– *transformation of the material* based on prior preparation of the waste for another use. The two most common logics are methanization and composting and the main uses are the production of energy, heat or use as an input in agriculture. Unlike recycling, the issue is not the homogeneity of the material resulting from the process but its possible use in relation to the three main uses mentioned above. ADEME defines composting as "a process of aerobic fermentation of fermentable materials under controlled conditions. It makes it possible to obtain a stabilized fertilizing material rich in wet compounds, which can be used, if of sufficient quality, as an organic amendment improving soil structure and fertility". It is a traditional form of recovery mainly targeting organic waste with two constraints: sufficient regularity of the process and non-toxicity of inputs. Methanization is "a treatment of waste or fermentable organic matter in the absence of oxygen in an anaerobic environment in tanks called digesters"[19]. Methanization has a double recovery in terms of organic matter (the "digestate") mainly used in agriculture and energy (biogas), which will in turn be recycled to produce energy and/or heat. This second aspect raises high expectations as an important element in reducing the ecological footprint. Cogeneration is defined as the use of by-products of a production process as fuel;

– *final disposal*, which is defined as "any operation which is not recovery even when the operation has as a secondary consequence the reclamation of substances or energy"[20]. This is the ultimate fate of waste. These are mainly landfill, burying or incineration. Landfills can be controlled, semi-controlled or so-called "wild". Landfilling requires the existence of natural and/or artificial cavities and raises the question of their control. Incineration raises the question of associated pollution.

The "reuse–recycling–transformation" (RRT) set is part of a waste hierarchy.

Waste recovery raises several questions: that of harmonizing the modalities of the socio-technical recovery chain, that of "legislation–regulation", that of implementing a circular economy and that of involving agents "in" and "around" the recovery chain. The reference to principles is then made: the principle of prevention (of waste production), the precautionary principle (anticipation of the negative effects associated with waste), the principle of coordination (of the elements of the recovery chain), the principle of coherence between the links, of collaboration, the principle of cooperation between the agents concerned "by" and "around" the recovery chain, the principle of hierarchical waste recovery, the principle of

19 *Ministère du Développement Durable, de l'Ecologie et de l'Energie*, www.developpement-durable.gouv.fr.

20 Directive 2008/98 of the European Parliament and of the Council of Europe of 19 November 2008, on waste.

producer responsibility of the "polluter pays" type, the principle of proximity (of waste treatment in relation to its place of production) and the principle of social and cultural inclusion of members of society.

11.6. Conclusion: arguing about boundaries

Conceptualizing waste means arguing about boundaries:

– "*life–death*": from an anthropological point of view, and in a first symbolic dimension, waste, when it crosses the boundary of the person who gets rid of it, constitutes a form of death, that of the product that we no longer want;

– "*clean–dirty*": from an anthropological point of view, waste is what marks the boundary of what is no longer considered clean, therefore of what becomes dirty. It is in this respect that a cultural definition of waste is possible, thus marking the possibility of comparing (and not being able to compare) the issue of waste in space and time. In the same vein, we will find the "*fresh–worn out*" tension of a chronological order, that of the "*keep–throw away*" of an anthropological order with all the symbolism of the rubbish that will then be attached to the waste from which it becomes necessary to get rid of;

– "*inside–outside*" of a methodological nature: when do we move from the inside to the outside?

– "*waste-free society–waste society*" of a socio-economic nature. This tension is *at least* representative of the "nature–culture" relationship as well as of the socio-technical trajectories followed. The theme of waste management is representative of a concept of a return to nature that will be found at the heart of the "linear –circular" tension. With regard to the relationship between waste and nature, we will also find the tension between "solid waste and organic waste", the two types of waste leading to a different approach to their management. The circular economy *stricto sensu* refers to an economic organization that takes into account the consumption of water, raw materials and energy sources. The aim is to close the lifecycle of products with regard to the objective of "zero waste";

– "*active–passive*" of an ontological nature depending on whether the waste is considered manageable or inevitable;

– "*high–low*" of a socio-political nature (where we find the ellipse "waste of class–class of waste");

– "*formal–informal*" of an epistemological nature, the crossing of the boundary tends to give waste an informal dimension (and we are then in the socio-political universe). It is the formal dimension that makes it possible to build a *problem- solving* approach, the informal one leading to an inductive approach (how to deal with it). It is also the institution that formalizes the formal dimension of waste, opening the field of what happens between the formal and the informal and allowing the logic of experimentation in the way of dealing with waste. It is also this tension that will make it possible to fix the cursor between the two dimensions, which are those of protection (of the environment of waste, for example) and conservation (of an area by refusing the entry of waste and by installing a systematic waste exit system).

– "*private good–public good*" of a political nature: when waste crosses the boundary of the private to become public, it is transformed from private good, it becomes public good;

– "*flow–stock*" of a kinematic nature, leading to the design of waste as a co-production or an externality that would then have to be dealt with, a question that is considered from an economic point of view. This boundary intersects with another, the one that is valid with the "*temporary–permanent*" duality of a chronological order. It is important to highlight here the existence of permanent global waste (see nuclear waste);

– "*supply–demand*" of an economic nature, that refers to the question of the existence of an economic opportunity for valorization. This boundary is the one that validates value chain reasoning from the industrial economy. The "supply–demand" tension also opens the door to the development of a waste trading activity. It is the value chain reasoning that will make it possible to specify steps that will link a collector and a pre-collection and/or collection activity, under the prism of recovery. The waste is then considered a convertible resource from the loss of value to a gain of value and we then cross-reference it with the tension, "useless–not useful–useful";

– "*linear–circular*", of a topological nature, the linear is an economy that considers the chain that operates between entry and exit, waste then being a co-product, a linked product or a "fatal product" resulting from the functioning of the chain and which is then reducible only with regard to the socio-technical dimension of the chain (see the theme of "zero waste" or that of "residue"). The circular economy emphasizes the movement that raises the question of opportunity throughout the circuit's operation. The "integration–disintegration" tension is related to the previous one, the integration of waste being representative of a circular design, while the disintegration of waste is related to a linear design;

– "*elimination–recovery*", of a practical nature, a tension that is in the shadow of the previous one and which is also of an economic nature. It is also an outcome of the relationship established between these two dimensions by public policies insofar as elimination (or not) as well as the trajectory of elimination is the issue of what is necessary or what is being discussed (and therefore what is at risk);

– "*north–south*", of a geographical nature, the movement of waste being marked by the trajectories of the flows. Their nature differs according to whether they are "south–south", "south–north", "north–south" or "north–north" circulations. Wasn't there a question of financing development from funding associated with waste transfers from north to south? This representation of flows is based on a geopolitical approach. The inhabitants of Dakar describe furniture and objects from Europe as "coming", bought and transported by containers to be resold on the spot. The inhabitants of Abidjan describe as "France, goodbye" vehicles of respectable age, whether they are cars, trucks or buses that have been recovered in Europe as part of "scrapping premiums" or because they have not passed the anti-pollution tests and are then sent to Africa;

– "*urban–rural*" of a human geography nature: urban waste is not the same as rural waste, its nature and reuse operating according to different logics both in terms of recovery (urban waste is a more complex technical and social process than rural waste, which can be just as dangerous as industrial waste – see pesticides) and distance (we hope for a closer and faster reuse of rural waste).

However, it is also important to highlight two aspects: waste is an element which, in the social modalities of its recovery, is a stigma of poverty as well as an important place for the development of the informal economy; it is also an element around which organized crime appears to proliferate, probably because society does not like to look at its waste, so it leaves it. It is also interesting to recall the theme of the Anthropocene, i.e. what qualifies the current age when human activity changes the state of the planet and is not characterized by extractivism. In this register, it is also referred to as "ultimate waste", which is the waste that is buried under the Earth's crust until technological progress makes it possible (or not!) to deal with it.

11.7. References

Bertolini, G. (1990). *Le marché des ordures : économie et gestion des déchets ménagers.* L'Harmattan, Paris.

Botta, H., Berdier, C. and Deleuil, J.-M. (2002). *Enjeux de la propreté urbaine.* Presses Polytechniques et Universitaires Romandes, Lausanne.

Chalmin, P. and Gaillochet, C. (2009). *From Waste to Resource.* Economica, Paris.

Douglas, M. (1966). *Purity and Danger: An Analysis of Concepts of Pollution and Taboo.* Praeger, New York.

Hoornweg, D. and Bhada-Tata, P. (2012). What a waste: A global review of solid waste management. Urban development series, knowledge papers, no. 15. World Bank, Washington, D.C.

Montsaingeon, B. (2017). *Homo Detritus : critique de la société du déchet*. Le Seuil, Paris.

Pesqueux, Y. (2011). Sustainable development: a vague and ambiguous "theory". In *Environmental Scanning and Sustainable Development*, Lesca, N. (ed.). pp. 25–48, ISTE Ltd, London and John Wiley & Sons, New York.

Monaco, A. and Prouzet, P. (2014). *Complexité du système océanique*. ISTE Editions, London.

de Rozario, P. and Pesqueux, Y. (2018). *Théorie des organisations*. Pearson, Paris.

Serres, M. (1990). *Le contrat naturel*. François Bourin, Paris.

Serres, M. (2008). *Le Mal propre : polluer pour s'approprier ?* Le Pommier, Paris.

Wilson, D., Velis, C. and Cheesman, C. (2006). Role of informal sector recycling in waste management in developing countries. *Habitat international*, vol. 30, issue 4, pp. 797–808.

12

When Fashion Brands Try to Adopt a Circular Economy

12.1. Introduction

The ecological balance of the textile sector is a disaster. It creates emissions equivalent to 1.2 billion tons of CO_2, more than all emissions from international maritime and air transport[1]. Another striking comparison is that each textile article can contribute 20 times its weight to greenhouse gases[2].

The textile industry has many environmental impacts (Bostrom and Micheletti 2016) at all stages of a product's lifecycle: raw material extraction, production, distribution, use and end-of-life (including recycling) (Box 12.1).

> What is the environmental impact of a simple T-shirt sold for less than $6 in the United States? This is the question that Pietra Rivoli asks herself in her book *The Travels of a T-shirt in the Global Economy* (2005). These environmental impacts are numerous. The production of raw materials, even natural ones, such as cotton, can be water-intensive and fertilizer-intensive and releases pesticides into the soil. Automated harvesting requires oil, as well as the transformation of the fiber into yarn, its cleaning and then its meshing. Dyeing uses polluting products such as heavy metals, which are then found in water, as well as in finishes applied to textiles. Cutting, making and printing patterns, wrapping the T-shirt in plastic and then cardboard, shelving and selling it involve energy costs. The simple use of a T-shirt involving repeated washing has an impact in terms of phosphates

Chapter written by Bénédicte BOURCIER-BÉQUAERT, Karen DELCHET-COCHET and Valérie FERNANDES.

1 https:///www.ellenmacarthurfoundation.org/publications/a-new-textiles-economy-redesigning-fashions-future, last accessed October 2019.

2 http://www.huffingtonpost.fr/2015/11/29/impact-textile-environnem_n_8663002.html.

polluting water. Finally, the end of the T-shirt's life is polluting, the degree of which depending on whether the product is buried, incinerated or recycled. In addition to these stages, transport also generates energy consumption at all stages. Thus, an ordinary T-shirt can follow the following geographical route: it leaves Texas where cotton is produced, then sent to China to be spun and garments shaped, then back to the United States, it is printed and sold. Once its buyer is tired, it is thrown away and then recovered to be redirected, most often to Africa, where it will be thrown away and then buried or incinerated.

Box 12.1. *Environmental impact of an ordinary textile (according to Rivoli 2007)*

Multiplying the environmental impacts, the quantity of clothing produced has doubled in 15 years (Ellen MacArthur Foundation 2017). However, 70% of clothes are not worn (or no longer worn), and each French person throws away an average of 12 kg of clothes per year[3]. This quantity of discarded clothing has increased by 8% in 1 year[4], as fashion too quickly condemns clothing that is still in good condition to waste (Joung 2013).

The situation is not only disastrous and absurd from an ecological point of view, it is becoming an economic challenge for the actors of the textile sector. Indeed, the textile industry is part of a context of increasing scarcity of natural resources and energy, exacerbated by overconsumption and the explosion of waste that needs to be managed.

A relatively recent phenomenon is that brands selling textile items (H&M, Intimissimi, Calzedonia, etc.), encourage consumers to bring back their old clothes (Bourcier-Béquaert *et al.* 2016). These initiatives are part of a specific regulatory framework in France: the management of textile waste has been promoted by the public authorities since 2009, in particular through the Extended Producer Responsibility (EPR). These regulations require manufacturers and distributors to collect and recycle as much of their end-of-life products as possible under environmentally friendly conditions. Another binding element for the textile industry in France is the draft law *"La loi anti-gaspillage pour une économie circulaire"* (Law against waste for a circular economy), which prohibits the destruction of unsold textiles by brands. However, beyond the question of the waste produced and the organization of recycling channels, should we not avoid producing waste and therefore consider the issue more broadly in a circular economy model? Is the regulatory and legislative arsenal a way to initiate circular practices within

3 www.ecogeste.fr.

4 Les Echos, June 23, 2017, *La filière textile donne une seconde vie aux vêtements*.

companies beyond waste management? Are these emerging practices expanding to other stages of the product lifecycle?

There is little research on textile recycling in a national ecosystem involving public actors, companies and consumers (Ekström and Salomonson 2014). This chapter proposes to focus on companies and their difficulties in implementing the circular economy, i.e. going beyond recycling practices alone. For this purpose, four brands, particularly those involved in the collection of used clothing, are studied.

12.2. State of play

12.2.1. *The circular economy: main principles and application in companies*

The principles of the circular economy

The CE (Circular Economy) is a rich reflection but still under construction. Prieto-Sandoval *et al.* (2018) reviewed no less than 162 academic articles on this subject. There is currently no shared definition (Khoronen *et al.* 2018; Murray *et al.* 2015). Behind the wide variety of definitions highlighted by Kirchherr *et al.* (2017), there is some convergence. The 4Rs seem to have a consensus: repair, reuse, repurposing and recycling:

– repair consists of extending the life of the product by making a part functional again, after the loss of its functionality;

– reuse aims to use a product second-hand by reintroducing it into the economic circuit;

– repurposing an object consists of its use for a purpose other than that for which it was intended from the beginning;

– finally, recycling is based on the use of parts or components of a product at the end of its life to manufacture others.

This foundation of the 4Rs of the circular economy must be complemented by three major points: the idea of a hierarchy in the levers used, the importance of eco-design thinking and the global nature of the CE.

First of all, some Rs have higher priority than others. The French law of August 17, 2015 on the energy transition for green growth *(Transition énergétique pour la croissance verte)* specifies that priority must be given to the reuse of products, then to their repurpose, then to their recycling and, ultimately, to their recovery.

Secondly, the CE integrates the way in which the product has been designed, conceived and produced through an eco-design approach. However, how do you repair a product that has not been designed to be repairable? How do you recycle non-recyclable materials? Eco-design thus seems to be the prerequisite for a successful circular economy approach.

Finally, the CE is a global approach. It implies low production and consumption of fossil energy and non-renewable raw materials. It requires the involvement of all stakeholders, particularly within a given territory. The CE set up by the organizations should therefore be able to address all these levers. This is a high target, and companies must start a real revolution in their production and marketing systems to achieve it.

Implementation of the circular economy by companies: a revolution to be achieved at all levels of the company

The transition from theory to enterprise implementation is particularly complex in CE. Indeed, the CE revolutionizes logistics flows within a company. Instead of being traditionally thought of in a linear way, via upstream and downstream flow management, circularity introduces the idea of a "closed loop supply chain" or "closed circuit" (Fulconis *et al.* 2016). As a result, in reverse logistics, flows flow back from consumers to manufacturers, creating a loop between the origin and the point of consumption of the product. In this approach, recycling takes an important place alongside unsold goods flows and repairs. The question of the point of sale is also important, as traditional and reverse logistics flows are closely related to it.

Circularity also implies a change in marketing approach. Indeed, since companies offer themselves as resource collection actors, customers/consumers also become their suppliers. The performance of the system depends on the consumer and their ability to take on a new role as a logistician, sorting what needs to be recycled or not, storing waste at home for a period of time and finally bringing it to the collection center (Anderson and Brodin 2005). Beyond their ability to sort and store clothes at home, consumers are central players because they decide between repair or replacement, they throw away, give away or return the product.

In companies, CE and its associated practices are therefore profoundly changing the design, supply chain (Fernandes and Kadio 2017), production, marketing and use and management of the end-of-life of the product. The transformation that the circular economy implies thus affects the forms of cooperation between actors (Lazzeri *et al.* 2017): consumers, customers and ordering companies and also suppliers, companies in the same sector of activity or located in the same territories and, finally, associative partners.

Which strategic issues within companies?

Faced with the revolution required by the implementation of the CE, companies are forced to ask themselves several questions. The first concerns the degree of implementation of the CE. Indeed, the CE is a goal to be achieved by organizations, but, in most cases, they do not succeed in activating all of its levers. Thus, there would be different degrees of implementation of the CE by companies. These levels of practice have not so far been studied extensively. Nevertheless, many questions arise. Does deploying an eco-design approach mean being in a circular economy perspective? Is recycling your waste enough? Will reselling second-hand products solve the problems? If all these actions can be carried out from a circular economy perspective, it is necessary to conduct a global and systemic reflection. This is the difficulty, for the time being, of these steps to establish the CE, each company itself defining the nature and scope of its CE actions.

The second question concerns the local nature of CE in a global industry. If the CE is considered as a model with no net environmental impact, restoring any damage related to the acquisition of the resource and parsimonious in terms of resources throughout the production process and product lifecycle, is this model compatible with a multinational's management methods? The course of Rivoli's T-shirt (2007) described in Box 12.1 raises doubts about this. The energy costs induced by the transport of goods and components should be controlled (Maillefert and Robert 2014). This is also true for the transport of used products for recycling. Companies involved in a CE approach should limit the geographical extension of sourcing and sales. This observation points to the likely difficulties of implementing the CE in a globalized context. Will companies restrict themselves geographically and return to more premises or will they adapt the CE to the context of a highly globalized world?

12.2.2. *CE in textile companies in France: a waste-based approach*

Textile companies in France

The textile industry in France is a significant sector. It includes about 2,400 companies in various fields such as the manufacture of yarns, fabrics and textiles for clothing, furniture or technical use. It represents more than 58,000 jobs and generated 13 billion euros in turnover in 2016 (Directorate General for Enterprise 2016; Union of Textile Industries 2017).

While there is no in-depth study on the circular economy within textile companies, in various respects, the textile sector appears specific (Belin-Munier and Moncef 2013). First of all, production is complex due to the very short lifecycle of products, which require a high degree of renewal. Distribution is strongly impacted by the limited capacity of the sales areas compared to the very wide range.

In addition, the entire supply chain appears to be globalized: production sites are far from sales outlets, resulting in import and export flows of both raw materials and finished products. The research by Belin-Munier and Moncef (2013) mentions few specific points concerning the sector in its post-consumer phase. These are returns for private sales, factory outlets, Internet sales or sales. The part relating to used products brought back by the consumer is not covered, nor is the recycling/re-employment/reuse of such products. This leaves the field open for exploratory investigations.

Context of textile EPR

Textile companies must orchestrate their circular practices in a specific context in France, at least as far as waste management is concerned. EPR (which is not European) has been in place since 2007. There is an eco-organization responsible for promoting CE in the textiles, household linen and footwear sector called ECO-TLC (the French words for textiles, household linen and footwear being "*textiles*", "*linge de maison*" and "*chaussures*", hence the acronym TLC). Created in 2009, its goal is the development of practices that promote the use of used or unsold products. ECO-TLC brings together many players, including marketers, citizens, collection operators (42,000 voluntary collection points in France), sorting operators (there are about 50 sorting centers in France) and recyclers. The latter restore value to used TLCs that cannot be used in their current state by transforming them into secondary materials that will be used for the manufacture of new products (wiping cloths, etc.).

Three types of loops have been defined by ECO-TLC (2016):

– a first short loop starts from the voluntary supply points (textile bins), then transport takes place to a sorting center which will identify products to be used as they are, which will then be put back into the loop when they are used. This is reuse. It concerns TLCs in good condition and that can be "used again for the same purpose for which they were designed" (according to the European Commission directive 2008/98/CE, Article 3);

– the second loop concerns the recycling and recovery of TLCs reinserted into the supply chain (closed loop);

– the third loop deals with recycling and recovery feeding raw material to other sectors (open loop). This is referred to as reuse when the materials can be used as rags or recycling for insulation in the building, for example.

In France, 62% of the textiles collected are reused, 31% are recycled, and the rest are sent for fuel and final disposal.

What difficulties do companies have in implementing the CE?

EPR encourages companies to collect used products, but to go further in the CE's approach, certain difficulties must be overcome.

The first difficulty lies in the necessary involvement of the consumer. They are responsible for the return of products to stores or collection points, as well as for the purchase of second-hand products (Hvass 2014).

Another more technical point: it is very complex to manufacture new clothes from used materials (Franco 2017). Indeed, recycling used textiles must make it possible to maintain the quality of the fibers, which is a real challenge. Modern textiles contain a lot of mixed fibers, so it is difficult to transform them back. Operators opt for "down-cycling", i.e. transformation into carpets or rags. This remark highlights the need to produce eco-designed clothing to make it easier to manage their end-of-life and to conduct research to preserve the quality of recycled fibers. Recycling and recovery actions benefit from these recent years of research and development, involving private actors, university research centers and associations. An example is the partnership between the H&M Foundation and the Hong Kong Textile and Clothing Research Institute (HKRITA), which has resulted in a technical solution for recycling mixed textiles.

Finally, the issue of suppliers and their involvement in CE practices is discussed by Lion *et al.* (2016). For example, the reuse of clothing leads sorting companies to export the least useful clothing in France to developing countries. Taking the case of Senegal, Bredeloup (2016) studies this outlet for collecting organizations, such as Le Relais, which has opened a sorting center near Dakar, for example. The author also questions the ethical aspects of these trade flows, in the sense that 20–30% of the clothes arriving in Africa are, in fact, not usable and become waste, not recycled and, therefore, polluting.

In short, the CE, proposing an ambitious approach, is difficult to implement in companies and requires a complete review of their operations. Textiles are no exception. Brands are engaged through an EPR in the collection phase. We still know little about their CE practices. Ekstrom and Salomon (2014) have shown the role of shops in the sector but do not study the sectors used by used textiles in more detail. We propose drawing up a table of company practices. The idea is to investigate whether the collection of old clothes in the shop is accompanied by other, less visible, CE devices. What role does it play in the CE systems of textile companies?

12.3. Methodology

In this exploratory study, we analyze the circular economy practices of four ready-to-wear brands that initiate collections of used textiles at the point of sale. Our study is being conducted in France, a country in which the CE approach is still emerging but is recognized as strategic. The EPR on textiles was implemented in 2009. Six years later, in 2015, the collection rate was only 33% and textile recycling was less than 30%[5]. The French law on the energy transition for green growth (*Transition énergétique pour la croissance verte*), promulgated on August 17, 2015 (article IV), highlights the CE. In December 2015, the European Commission, by voting for the "circular economy package", decided to allocate 650 million euros under the Horizon 2020 program and 5.5 billion euros under the French Strategic Investment Fund (FSI). Since then, practices have been on the rise (+8% in terms of collection in 2017) but remain far behind other countries, such as Germany and Sweden.

The selected brands are known and have collected used textiles over the past 12 months. We have chosen four brands that are well established in France: Zara, H&M, Cyrillus and Camaieu. These brands have different characteristics with regard to several key collection criteria (Table 12.1):

– different degrees of internationalization: H&M and Zara are totally globalized brands and Cyrillus and Camaieu are less international;

– the size of their distribution network: the distribution of Cyrillus and Camaieu products is less significant;

– the implementation date of the collection systems: this was spread between 2013 (H&M) and 2017 (Zara) and is not at the same level of maturity and deployment in all companies.

	H&M	ZARA	CYRILLUS	CAMAIEU
Characteristics of the brand				
Nationality	Swedish	Spanish	French	French
Degree of internationalization (high vs. low)	High (4700 stores in 70 countries)	High (1900 stores in 89 countries)	Low (52 stores in France and 7 abroad)	Low (650 stores in France and 250 in 15 countries)
Characteristics of the collection				
Name of the program	*Bring it/Conscious*	*Join life*	*Je recycle* (I recycle)	*Et si on recyclait vos vêtements?* (And if we recycle our clothes?)

5 http://www.ecotlc.fr/ressources/RA_Eco_TLC_2015_web.pdf.

Starting year	2013	2017	2016	2015
Collection of information	55,000 T	N.C.	70 T	40 T

Table 12.1. *Presentation of the brands studied*

Our analysis is based on data available at the point of sale, on the brands' websites and through their communication. Annual reports were reviewed when available. In-store observations and simulations of customers wishing to deposit used textiles were carried out. Quick interviews were conducted with sales staff to verify the information collected online. The CSR managers were contacted, and we were able to conduct an in-depth interview with the H&M manager in October 2017. Finally, reading press articles helped us to supplement these data.

The objective is to identify, for each brand practicing the collections, 1) the methods of collection (organization, systematic nature, globalized scale or not), 2) the characterization of the loop, including this waste management, and 3) the inclusion of this practice in a business strategy.

12.4. Results

12.4.1. *There is a collector and... a collector*

First of all, the collection operations observed reflect a significant investment by the company through the marking of the collection store, its layout via furniture and the training of salespeople in this type of operation.

These collection operations show different degrees of sustainability. Some companies have set up operations on a permanent basis (Zara, H&M, Cyrillus). Their purpose is to cover all their shops. Thus, H&M deploys its collection worldwide. Zara is currently deploying the scheme: not all stores are yet collecting. On the other hand, Camaieu only carries out an annual operation of 10 days (Table 12.2). To guarantee a large collection, several private labels, except Zara, offer the consumer a reward when returning their old clothes (a discount or discount voucher to be used on a future purchase). The conditions for benefiting from these rewards are not very restrictive: they exclude few items, do not place conditions on the brands collected and set very accessible thresholds for benefiting from the discount (three items reported).

Another observation is that Zara offers a new type of collection that other brands do not yet offer: home collection that can be done when an online order is delivered. This system exists in Spain but is not yet deployed in France. Collection should therefore be adapted to the multi-channel profile of the brands with all the logistical organization that this entails.

The collection follows two different organizational modes in terms of partners. In the first case, the brand uses a single partner (I: Co) present in 60 countries, which is responsible for all operations (transport from drop-off points to a sorting center, sorting, recycling or routing to new destinations for reuse or processing). I: Co's warehouses are highly centralized. There is only one in Europe, which means that the cost, including environmental costs, of moving textiles from collection points to sorting centers is very high. On the other hand, I: Co has efficient solutions for textile recycling. In the second case, the brand collaborates with many partners. These are associations and have a charitable scope (in particular, the integration of the most disadvantaged to whom they bring work). They are often local and are located in the heart of the collection areas. There are therefore two waste management circuits, one multinational and the other more local. In the latter case, the brand's commitment is stronger since logistics can be shared. Zara, for example, transports the collected used textiles to the sorting centers.

12.4.2. *A still partial implementation of the CE*

The analysis of broader circular practices and their inclusion in the CSR strategy of brand names leads us to make several observations (Table 12.3).

H&M and Zara, ready-to-wear giants, have a CSR approach and claim to have integrated CE into their strategy. These brand names play an active role in the industry by participating in research projects on new recycling technologies. The positions of Cyrillus and Camaieu are somewhat different. Cyrillus claims a strategic commitment to the environment but not directly on the subject of waste and the CE. As for Camaieu, the strategic axes are not communicated.

Apart from Camaieu, which communicates relatively little about point-of-sale collection, companies include this approach in a more global CSR approach. They opt for complementary social actions and choose partners to support the integration of disadvantaged people at Cyrillus and Zara. They also make financial donations on behalf of Zara and H&M.

	H&M	ZARA	CYRILLUS	CAMAIEU
Temporary/permanent	Permanent	Permanent	Permanent since 2018	Temporary (10 days/year)
Perimeter (physical/Web)	Physical: all H&M stores worldwide	Physical: some stores Web: Pick-up at home when ordering in Spain	Physical: all boutiques outside Clermont-Ferrand and Galeries Lafayette	Physical: all stores in France and directly operated stores in Europe
Products/brands collected	All textiles All brands Clean and dry textiles Whatever their condition	All textiles All brands Clean and dry textiles Whatever their condition	Clothing and household linen All brands Clean and dry textiles Exclusion: shoes, leather accessories, duvets, pillows and lingerie	Women's clothing and shoes All brands Clean and dry textiles Exclusion: jewelry, linen, accessories, bags and household linen
Point of sale	Presence of POSD[6] and POSI Information/training of salespeople	Presence of POSD and POSI Information/training of salespeople	Presence of POSD and POSI Information/training of salespeople	Presence of POSD and POSI Information/training of salespeople
Bonus	For each bag of at least three pieces, -15% on an item	No remuneration	1 bag deposited, 1 coupon of €5	20% discount on the preferred product
Conditions of the offer	Two vouchers per person per day		Up to three bags (of three textiles minimum)/person/day	One discount per day and per person
Circular economy partner(s)	*I: CO* Single global partner (any solution, global)	*Le Relais*, *Caritas*, *Red Cross*, *Salvation Army*, *CEPF* and *Redress* Multiple partners, variable/country, Social Economy Organization (SEO), cooperation of different actors at a local level	*Le Relais*, *Emmaüs France*, *Emmaüs Alternatives*, *Sock in stock*, *Apivet*, etc. Multiple partners, SEO, national or even regional sites	*Caritas* and *Weaving Solidarity* and then *I: OC* in 2018 Passage from a local partner from SEO to a single global partner (any solution, global)

Table 12.2. *Description of in-store collection operations by brand names*

6 POSD: Point of Sale Display; POSI: Point of Sale Information.

	H&M	ZARA	CYRILLUS	CAMAIEU
Strategic axis CSR	"*Recycle your clothes*" is one of the three commitments in terms of sustainable development	"*Recycling and efficient use of resources*" is one of the seven strategic priorities	Respect for the planet is one of the three commitments	N.C.
Reuse	Donations to associations	Donations to associations	Second-hand site[7]/donation to associations	Partner of *Tissons la Solidarité* (repair and resale in second-hand clothing) in 2017 Since 2018, given the partnership with I: CO Reuse and Recycle
Reuse and repurpose	Textile products not suitable for use in clothing to create other products (rags)	Articles made of 100% cotton, wool or polyester to make new fabrics "*Shape the Invisible*" collection not marketed (work of creators reusing old clothes)	No specific action mentioned but: 61% reuse 36% recycling 3% waste N.B.: Extension of the "second-hand" approach to books and school bags	
Recycling	Automotive industry, construction	Automotive industry, construction		
Marketing of "green" ranges	Organic cotton or recycled materials	Organic/recycled fibers, water and CO_2 parsimony	Organic cotton for babies and children	No
Research/recycling	Chemical recycling with *Kering* and *Worn Again*	Work with *Lenzing* and *MIT* on innovative projects	N.C.	N.C.
Societal actions related to collection	1 kilo of textile collected = 0.2 E donated to *UNICEF* France	Donation to CARITAS of €3.5 million for collection points	Non-financial support (social inclusion)	

Table 12.3. *Description of circular economy actions of brand names*

7 https://www.secondehistoire.fr/fr/.

Thanks to their partners, the brands are positioned on the three stages of end-of-life management: Reuse, Repurpose and Recycle. The supply of new products resulting from this recycling, i.e. recycled fabrics, reflecting a form of circularity, is fairly rare in the companies analyzed because it concerns only a small number of products. It is even totally absent from Camaieu. It should be noted that the products are more widely presented as responsible, i.e. they are made from fibers that can be either recycled or organic. H&M, with its Conscious range, and Zara, with Join Life, present responsible product ranges in organic and recycled fibers. Cyrillus, for its part, offers a Capsule collection for babies and children exclusively in organic cotton, stating that its objective is to create a collection made of recycled fibers. But the share of brands in the turnover and in the number of references offered is quite limited.

12.5. The limits of the actions implemented

The study of the four cases shows that the brands are developing a "new business": that of a waste collector. What are the consequences of this new profession in terms of CE?

The first remark is to point out that this job as a collector does not necessarily lead them to engage in all the downstream operations of waste management. These are often delegated to third party companies. The commitment to waste management as a circular resource is therefore still limited.

As a second remark, the brands studied, thanks to their collection partners, are positioned on the three aspects of Reuse, Repurpose and Recycle (Table 12.3). However, the French "*Transition énergétique pour la croissance verte*" (Energy transition for green growth) act emphasizes that there is an order of priority in the options considered:

> the prevention of waste production, in particular through the reuse of products and following the hierarchy of waste treatment methods, the reuse, recycling, or where possible, repurposing of waste.

The idea is above all to prevent the production of waste and not to collect and recycle as the brand names do. One of the levers to prevent waste production is to increase the functional and environmental quality of products through eco-design. On this subject, very little information is provided to us. Brands do not seem to be developing initiatives to improve product quality, which is a guarantee of an increase in shelf life before recycling. It should be noted, however, that recently, to increase the lifespan of their products, H&M has been providing washing advice to its customers.

The recycling capacity will also depend on the design. Indeed, current technologies do not allow all textiles to be recycled. According to the Zara site, "today, technology only allows the textile recycling of garments that are 100% cotton, wool or polyester". Brands then have two options: either develop research programs to find technological solutions for fiber recycling or use only recyclable fibers to produce their clothing and be part of an eco-design process, from a CE perspective.

Avoiding waste production means promoting reuse. This question of reuse is interesting because it can occur before and after collection: upstream, via second-hand sales sites, or downstream, via collection and donation to associations. With regards to the upstream phase, several practices exist. For several years now, we have been assisting in the development of online sales sites for second-hand clothing offered directly by brands such as Petit Bateau and Jacadi, for example. Other brands will partner with an online store, like Patagonia with eBay. Cyrillus pays particular attention to reuse by being located at two levels: upstream, via the site created by the brand for reselling second-hand products, and downstream, via the donation to associations of collected clothing, some of which are in perfect condition. The latter come from clothing collected in stores. But we do not have any figures as to the percentage of these products actually reused. Our information only covers the following: on non-recoverable clothing, *61% reuse, 36% recycling and 3% waste*. In general, brands should organize comprehensive information on the actual fate of the clothing collected.

A last axis concerns the extension of the product's useful life through repair. This subject is not addressed by the brands we have analyzed. This does not mean that the products have never been repaired: it is still common for consumers to repair a zipper, sew a button or fix a snag. However, this repair is not orchestrated by the brands studied. Nevertheless, some of them offer to repair their products such as Patagonia or The North Face. In addition, some recycling partners, such as Emmaus Alternatives, repair clothing that can be damaged after Cyrillus has collected and sorted it.

The practices of the brands are therefore varied and complementary. Nevertheless, taking environmental impacts into account from the design stage of the product makes it possible to resolve certain issues related to its end-of-life. In other words, the prospect of increased lifespan, reuse, repurposing and recycling requires companies to think differently about their products.

Another limitation of these collection systems is that these operations are ambivalent with regard to the CE. From a marketing point of view, this type of collection operation works to build customer loyalty by encouraging customers to consume again, once their old clothes are returned. This is precisely against the establishment of a circular economy, since consumption is fueled by recycling. More broadly, it is the very model of fast-fashion, that must be challenged when implementing circular practices. Brands do not commit themselves to a positioning more focused on basic collections than on fashion. However, fashion is an instrument of planned obsolescence since it disqualifies clothes, not because of their condition, but simply because of their cut, material or color. As a result, what is the credibility of collection actions when they are carried out in an industry that promotes unsustainable consumption practices such as fast-fashion?

12.6. Conclusion

Our exploratory research provides an overview of the main practices of circular economy by ready-to-wear companies initiating the collection of old clothing. We are dealing with more or less complete approaches to circularity. It seems that the largest firms are also the ones that offer the most numerous approaches, including *sourcing* and technological research to improve recycling. This observation should be seen in the context of the considerable human and financial resources required to transform a company and lead it to circular practices.

Our results highlight a number of contradictions that companies will need to resolve in order to adopt circular practices. One of our main results is to show that we are dealing with different approaches to circularity: global management by the multinational company H&M, national management by the multinational company Zara, regional management by Cyrillus. These different approaches have an impact on logistics management, in particular. The international management of circularity contradicts the very notion of a circular economy that promotes local, parsimonious management in transport (mileage and CO_2 emissions). This refers to a research approach designed to deepen the relationship between the circular economy and the notion of geographical proximity.

Moreover, it seems difficult for fast-fashion brands to promote circularity since it helps to accelerate consumption. In the options for managing the flow of collected used textiles, the second-hand, offered on behalf of the brand, is infrequent. However, it would allow the brands that practice it to move away from their fast-fashion image. Indeed, it testifies to the durability of clothing. This has consequences in the marketing field that still need to be explored. Thus, the slow-fashion movement, which appeared in the 2010s and was strongly driven by the Rana Plaza disaster (Chapuis and Pündrich, 2016), seems, to us, to be an interesting avenue to explore, because it tries to integrate the framework of the circular economy from the eco-design phase, by rethinking the economic model of companies.

Of an exploratory nature, our study does not provide a complete picture of the implementation of circular practices by these companies, which nevertheless initiate the collection of used textiles. It will be interesting to continue this research based on interviews with all the actors involved in this approach: ready-to-wear companies, consumers and recycling partners from both the SSE and the private sector.

12.7. References

Anderson, H. and Brodin, M.H. (2005). The consumer's changing role: The case of recycling. *Management of Environmental Quality: An International Journal*, 16(1), pp. 77–86.

Belin-Munier, C. and Moncef, B. (2013). Les chaines logistiques multi-acteurs internationales dans le textile : le point de vue des acheteurs. *Logistique & Management*, 21(34), pp. 59–70.

Bostrom, M. and Micheletti, M. (2016). Introducing the sustainability challenge of textiles and clothing. *Journal of Consumer Policy*, 39, pp. 367–375.

Bourcier-Béquaert, B., Damay, C., Delécolle, T. and Loussaïef, L. (2016). Collecte de produits usagés en points de vente : de nouveaux rôles clients-enseignes. *Logistique & Management*, 24(1), pp. 43–56.

Bredeloup, S. (2016). FEGG JAAY : fripe business ou fripe éthique au Sénégal ? *Mouvements*, 87, pp. 142–154.

Chapuis, S.M. and Pündrich, A.P. (2016). DELLA en mode slow fashion. *Revue Recherche et Cas en Sciences de Gestion,* (16), pp. 7–19.

Direction générale des entreprises (2016). L'industrie textile. [Online]. Available at: https://www.entreprises.gouv.fr/secteurs-professionnels/textile-mode-et-luxe [Accessed 22/02/2018].

ECO-TLC. (2016). Rapport d'activité. Activity report. [Online]. Available at: http://www.ecotlc.fr/page-307-rapports-d-activite.html. [Accessed 22nd February 2018].

Ekstrom, K.M. and Salomonsonn N. (2014). Reuse and recycling of clothing and textiles – A network approach. *Journal of Macromarketing*, 34(3), pp. 383–399.

Ellen MacArthur Foundation (2017). A new textiles economy: Redesigning fashion's future. Report. [Online]. Available at: https://www.ellenmacarthurfoundation.org/publications [Accessed 29th October 2019].

Fernandes, V. and Kadio, C. (2017). Intégration de l'économie circulaire dans le management de la supply chain : une étude exploratoire. *Logistique & Management,* 26(1), 15–25.

Franco, M.A. (2017). Circular economy at the micro level: A dynamic view of incumbents' struggles and challenges in the textile industry. *Journal of Cleaner Production*, 168, pp. 833–845.

Fulconis, F., Paché, G. and Reynaud, E. (2016). Vers une nouvelle forme de croissance économique. *Revue Française de Gestion*, 261, pp. 128–149.

Hvass, K.K. (2014). Post-retail responsibility of garments – A fashion industry perspective. *Journal of Fashion Marketing and Management*, 18(4), pp. 413–430.

Joung, H.M. (2013). Materialism and clothing post-purchase behaviors. *Journal of Consumer Marketing*, 30(6), pp. 530–537.

Lazzeri, Y., Bonet, D. and Domeizel, M. (2017). *Économie circulaire et territoires*. PUP, Aix en Provence.

Lion, A., Macchion, L., Danese, P. and Vinelli A. (2016). Sustainability approaches within the fashion industry: The supplier perspective. *Supply Chain Forum: International Journal*, 17(2), pp. 95–108.

Masson, S. and Petiot, R. (2012). Attractivité territoriale, infrastructures logistiques et développement durable. *Cahiers Scientifiques du Transport*, 61, pp. 63–90.

Maillefert, M. and Robert, I. (2014). Écologie industrielle, économie de la fonctionnalité, entreprises et territoires : vers de nouveaux modèles productifs et organisationnels ? *Développement durable et territoires*, 5(1).

Murray, A., Skene, K. and Haynes, K. (2015). The circular economy: An interdisciplinary exploration of the concept and application in a global context. *Journal of Business Ethics*, 140, pp. 369–380.

Pietra, R. (2005). *The Travels of a T-shirt in the Global Economy*. John Wiley & Sons, New York. Available at: https://books.google.co.uk/books/about/The_Travels_of_a_T_Shirt_in_the_Global_E.html?id=8l86BAAAQBAJ&printsec=frontcover&source=kp_read_button&redir_esc=y#v=onepage&q&f=false.

Prieto-Sandoval, V., Jaca, C. and Ormazabal, M. (2018). Towards a consensus on the circular economy. *Journal of Cleaner Production*, 179, pp. 605–615.

Union des industries textiles (2017). Rapport d'activité 2016-2017. Activity report. [Online]. Available at: www.textile.fr [Accessed 22nd June 2019].

13

The Circular Economy and Packaging: Challenges and Avenues for Reflection

In a circular economy approach, packaging deserves special attention because the stakes are so high and their impacts so significant.

The stakes are direct: they are linked to the production of packaging, its destruction or not, its recycling or not, etc. The figures are quite clear: packaging accounts for 40% of global plastic production (Geyer *et al.* 2017). However, many of these plastics are designed to be discarded after only one use. The immediate consequence is measured in terms of pollution: almost half of the plastics already produced become waste in less than three years, and therefore more than 75% are now waste. Between 1950 and 2015, we generated 8.3 billion tons of plastics and 6.3 billion tons of plastic waste, a figure that could almost double if we do not change a trajectory that has become unsustainable, as Caroline Janvier, a Member of Parliament, reminded the *Assemblée nationale* on July 11, 2019[1].

Thus, in 2012, each French national generated 277 kg of household waste (excluding bulky waste), 1/3 of which was in packaging, which is perceived as ephemeral, ancillary, intended for disposal or destruction[2], etc. According to the same source, the packaging sector in France in 2018 represented nearly 35 billion euros in turnover and 200,000 direct jobs.

The stakes are also indirect because the functions attributed to packaging are multiplying and growing in importance due to economic developments (mass

Chapter written by François CABARET.

1 http://www.assemblee-nationale.fr/15/europe/rap-info/i2132.asp.

2 2016 Ademe study cited by the *Conseil National de l'Emballage* (French National Packaging Council) in its May 2018 report "*L'emballage en France*".

production of finished products, even in the food sector, international trade, etc.) and sociological developments (changes in lifestyles, "Out-of-home" consumption, self-service or remote purchases, etc.). As Adetem, the French national association of marketing professionals, already analyzed in 2004: "the increase in the amount of packaging followed that of self-service" (Pinet 2004). In a rather abrupt short cut, we could say that fast food, self-service and e-commerce have boosted packaging consumption.

It is very revealing in this respect that, historically, the first approach of EPR (Extended Producer Responsibility) included in the legislative texts in the autumn of 2019 under discussion in the Assembly and the Senate was for the packaging sector. It has made it possible to create an "eco-packaging tax", to finance companies in charge of collecting and recycling packaging, such as "*Eco-Emballages*" ("eco-packaging"). This, on its website, provides a reminder of the history presented in the box below.

1972: Creation of the "Polluter–Payer" principle.

As early as 1972, OECD member countries decided on the principle that economic actors should take responsibility for the pollution generated by their activities.

1991: Riboud and Beffa reports.

Antoine Riboud, President of BSN (former name of Danone), and Jean-Louis Beffa, President of Saint-Gobain, submit their report to the Minister of the Environment Brice Lalonde, in which they define the principles for dealing with "lost packaging" as part of the Extended Producer Responsibility (EPR).

1992: Creation of Eco-Emballages.

A non-profit company, Eco-Emballages was created to set up a selective collection system that allows consumers to give a second life to their packaging waste. They also support packaging reduction actions.

1993: First national communication campaign on sorting. Its slogan: "sorted packaging is no longer lost packaging".

Box 13.1. *History of "Eco-Emballages"*

The emblematic product of this packaging sector is, without a doubt, the Bottle, which immediately raises the question of Consignment. The initial draft of the "waste for a circular economy" bill provided for the introduction of a "deposit for reuse, repurposing or recycling" system for products consumed by households, such as beverage packaging. The Senate Committee on Sustainable Development decided to remove the recycling deposit and only keep the deposit for reuse or repurposing. They believe that the recycling deposit is aimed, above all, at plastic bottles, which will thus be perpetuated[3], and that the choice of material: glass, plastic, cardboard, etc. will therefore be the most important.

As early as 1938, a law (published in the *journal officiel* of January 13, 1938) required brasseries to return sparkling water bottles for deposit, but this practice of returning glass bottles for reuse was gradually abandoned from the 1970s onwards in favor of disposables, under pressure from operators in the sector, distributors and others.

Better packaging design therefore directly aims to better control its proliferation and, if possible, to promote its reuse or even recycling. This indirectly leads to the fight against the waste of the products contained, or even their ecological footprint (transport, storage, etc.).

The decisions to be taken when designing packaging are numerous and their consequences complex, whereas for a Product Manager, by nature, it is a secondary, or even ancillary, subject compared to the decisions to be taken when designing the product itself! To help a more ecologically-responsible reflection on packaging, we will start from the functions fulfilled by it, which we group into four themes:

– division function;

– protection functions;

– marketing and communication functions;

– service functions.

13.1. Division function

Defining packaging means defining the quantity of product it will contain – its "division". This choice is not neutral since it will, of course, determine the price of the final product as it will appear to the buyer, but will also influence the level of consumption, the shelf life once opened. This choice can therefore be a waste of the contents. It also leads to the multiplication of atypical packaging formats, incompatible with logistics standards, cartons, pallets, racks, etc. resulting in additional logistics costs.

Faced with the complexity of the problem, several avenues are being explored.

3 AFP press release dated September 24, 2019.

13.1.1. *Bulk goods*

This consists of organizing, at a retail outlet scale, whether self-service or not, the "bulk" sale of products, the argument being that only the quantity actually needed is then purchased, and – in addition – only standard bags or sachets (or even an existing container brought by the customer) are required. Thus, we have seen an increase in the number of experiments, 80% for food products such as wine, cereals and confectionery, with traditional grocery stores and non-sedentary sales in our markets, and 20% for hygiene or cosmetic products, even medicine[4].

Since the 2000s, the development of specialized bulk sales shops or departments has seen two successive waves. The first has a "first price" positioning, focusing mainly on the savings generated by more precise consumption and lower transport and packaging costs. The results of this first wave seem to be quite mixed, especially in the mass retail sector. It has remained rather discreet on the results obtained, in particular, concerning markdowns (losses, theft, etc.) and therefore, paradoxically, waste, as well as on additional maintenance costs.

A more recent wave claims a more "premium" positioning, emphasizing a more global concept – locavore, organic, eco-responsible, etc. – for a more militant target of consumers, as this testimony proves[5]:

> Sourcing from a bulk grocery store is "one of my New Year's resolutions: I became a vegan last year and I realized that it was to achieve zero waste", says Nolween Delage, a 25-year-old graphic designer and customer of La Recharge in Bordeaux, AFP. This grocery store has reintroduced the deposit system with the objective of working in a short circuit with local producers, explains its co-manager, Jules Rivet, to AFP, whose clientele has quadrupled in four years. A figure that comes as no surprise to Célia Rennesson, Executive Director of Réseau Vrac, the organization that brings together 600 players in the sector: "In fact and in figures, France is ahead of its European counterparts," she told AFP. Thus, from 15 grocery stores specializing in bulk a few years ago, France now has 200, far ahead of Belgium, Germany and Great Britain. And according to Credoc (*Centre de recherche pour l'étude et l'observation des conditions de vie*), 47% of French people bought food products in bulk at least once in 2018, compared to 32% in 1998.

4 Célia Rennesson, Executive Director of Réseau Vrac, the organization that brings together 600 stakeholders in the sector, AFP on March 23, 2019.

5 The French are increasingly fond of bulk sales, published by *Par Sciences et Avenir* with AFP on March 23, 2019 at 12h00.

The success of these new concepts is probably due to the ecological benefit, perceived as superior to the disadvantages and risks related to hygiene, as the products are unprotected, the more difficult traceability of the products, the maintenance of the department or store, generating additional costs, etc.

13.1.2. *The sale in unit packaging*

A variant of bulk, but seeking to provide solutions to problems of hygiene and traceability of bulk products, unit packaging could be of interest to the pharmaceutical sector, or the snacking and fast food market.

With regard to medicine, for example, several countries (e.g. Scandinavian countries) or sectors (e.g. hospitals) have opted for "bulk" sales. The pharmacist delivers the exact quantity stipulated by the prescription, in order to avoid any waste (in France, about half of prescription drugs, 40% more than our neighbors, will never be consumed). This "bulk" solution does not meet with the support of pharmacists in both "cities" and hospitals, who highlight both hygiene problems and the risk of errors when taking these drugs, hence the idea of providing unit packaging (in cellophane or other packaging, or attached on a string (like rosaries or candy necklaces)) in which the exact quantity needed would be distributed.

The principle could be adopted for snack food products, hygiene and cosmetic products, even diapers, etc. ("bodycare") or detergent products, for the home, laundry, etc. ("homecare"). In such cases, any additional packaging costs would be compensated by reducing waste.

13.1.3. *Sales in the so-called family or "giant" promotional packaging*

On the other hand, the trend towards increasing the size of packaging aims – *a priori* – at certain economies of scale, by spreading some of the costs (labor, processing, etc.) over larger quantities of product. However, they are more questionable in terms of waste.

13.2. Protection function

The initial or "first-pack" container is the one in contact with the contents, often supplemented by an on-pack container, then different boxes, crates, etc. according to the logistical needs.

The choice of material used for the container must first respect the compatibility between the contents and the container with which it is in direct contact. In short, it is the protection of the container against certain "aggressions" of the content! Against the risks of oxidation, corrosion, chemical reactions, etc., the neutrality of the material used is essential, for example, that of glass in food or cosmetics is very good. Conversely, the container must also protect the content from external chemical aggressions, oxidation, UV, odors, etc. It thus contributes to the preservation of the product.

A first major historical step was taken in the 19th Century by Nicolas Appert. This French inventor was the first to develop a method of preserving food by subjecting it to heat in hermetic and sterile containers (jars, welded metal cans, etc.): the famous tin can.

More recently, the invention of the complex film Tetra Pak, in 1951, by the Swedish industrialist Ruben Rausing, combines three layers: cardboard, plastic and aluminum, thus optimizing the quality of this new material. Made of a continuous tube, it can be filled and sealed under vacuum, guaranteeing the long preservation of the liquid contained.

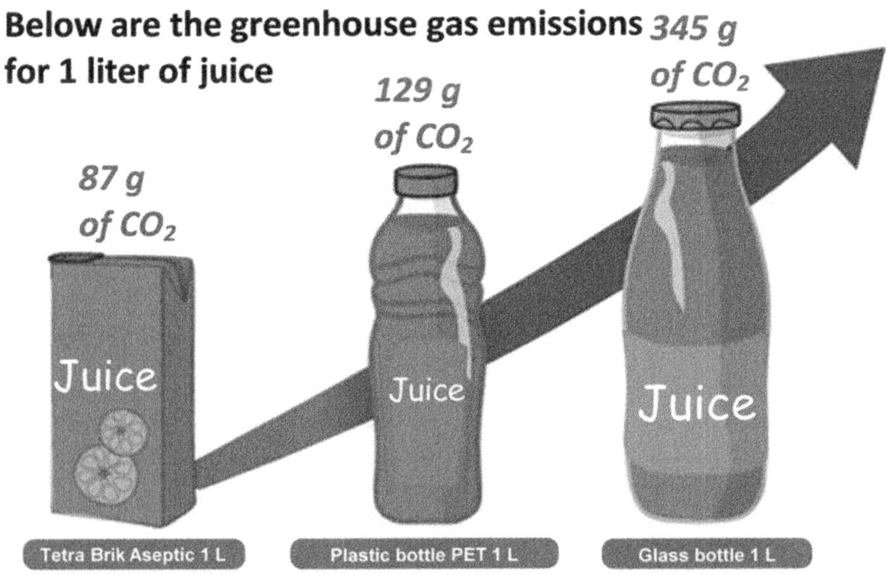

Figure 13.1. *Greenhouse gas emissions per 1 L of juice, Lifecycle Assessment Bio Intelligence Service Study, March 2008. For a color version of this figure, see iste.co.uk/delchet/circular.zip*

Today, some detractors reproach the difficult recycling of this complex film, in particular, the separation of its three components – especially the plastic and aluminum. But its advocates point out other advantages: less energy needed to produce it, compared to glass or PET bottles, less logistic costs (transport, storage outside the cold chain and therefore, again, less wasted energy). The total comparative assessment carried out as part of the LCA (Lifecycle Assessment), comparing the three containers (Figure 13.1) gives food for thought, and everyone innovates to reduce the defects of their product!

Two major phenomena in distribution have "exploded" these expected protective functions of packaging:

– In the 20th Century, the generalization of self-service sales, which today concern the vast majority of food products and a large part of personal and household equipment. Products "left to their own devices" must be protected from breakage, impact, theft, hence, for example, large rigid packaging for small products that are too easy to conceal (batteries, lighters, etc.). Against the "violation" of the product, systems of tearable guarantee strips, heat-sealed caps or heat-shrink films (known as "skin-packs") have been added to ensure that the packaging has not been opened, that the product has not been touched, soiled, etc.

– In the 21st Century, the emergence of "e-commerce", one of the main difficulties of which lies in the supply chain, particularly at two levels. First of all, the first link in this chain, order picking, requires boxes, cartons, overpacking of all shapes and sizes. The last link in this same chain, delivery to homes or collection points, considerably limiting any grouping, rationalized transport on pallets, etc. also contributes to the proliferation of packaging, often quasi-individual. The development of e-commerce calls into question the traditional logic of the different levels of packaging: primary, secondary, tertiary, which should be completely rethought (Fulconis and Philipp 2018). As an illustration, the promotional operation of the Chinese merchant site Alibaba, its "Single day", on November 11, 2019 generated, in that one day, 1390 million packages to be delivered!

In addition, in e-commerce, the empty space, for some products, is more than 60% of the volume of packaging[6].

More generally, the development of international trade and "delocalized" production contribute greatly to the multiplication of secondary and even tertiary packaging, cartons, crates, pallets, containers, etc., even if extraordinary efforts at

6 Study published by Forbes Insight in collaboration with DSSmith, one of the leading European packaging players. Available at: https://www.dssmith.com/fr/packaging/strateges/livres-blancs/leconomie-de-lespace-vide.

standardization have been made, thanks in particular to the container, boats still transport empty space for ¼ of their space[7].

13.3. Marketing and communication functions

Marketing can use packaging as a tool for product recognition and differentiation, such as the round bottle of Orangina or the curved Coca Cola bottle, which will increase the number of bottle shapes and their sizes and reduce the possibilities of standardization.

It has also been shown that the perception of the product is influenced by the nature and shape of its packaging, by a "sensation transfer". Thus, according to Herbreteau and Jacquot (2017), there is a different perception of the taste of the same yogurt depending on whether it is in a square or round pot. The same marketing observation applies to the perception of a perfume and the perception of its bottle (shape, material).

To "blur the lines" as to the public price of the product (psychological price or comparison with that of the competition), marketing can vary the quantities sold (which are supposed to be rounded to whole decimals): 205 gram chocolate bar, or 170 grams, 70 or 60 centiliter bottles, 80 gram cosmetic cream jars, etc. These actions on quantities lead to additional manufacturing and logistic costs.

A more embarrassing trend has been the temptation to use so-called "thieving" packaging. These are boxes or jars whose apparent external volume is misleading about the idea of the volume actually contained, such as double-walled jars with very thick walls, with a "bell-shaped" bottom. This approach is reprehensible if there is a desire to deceive, and which – incidentally – results in additional costs and consumption of packaging and materials, etc.

Packaging is also a communication medium: it must include legal information: quantity, composition, origin, dates of manufacture, sales limits for fresh products. The multiplication of secondary packaging (cardboard box containing a tube of toothpaste, for example) is explained by the amount of marketing information that we want to communicate: promotion of the product, its effectiveness, advice on use or implementation, hence also the leaflets, range presentations, etc., not to mention the essential "bar code" that we print there for the traceability of the product, its passage in crates. This communication is not – in itself – reprehensible. But what

7 Study published by Forbes Insight in collaboration with DSSmith, one of the leading European packaging players. Available at: https://www.dssmith.com/fr/packaging/strateges/livres-blancs/leconomie-de-lespace-vide.

raises questions is its impact on the proliferation of packaging and its difficult recycling.

This phenomenon is further aggravated by self-service sales (already mentioned), because packaging fulfils a "merchandizing" function, in the departments where it is found to be "the silent seller" of the product, according to a dedicated formula. It must attract the customer's attention, arouse their curiosity, their interest and make them want to buy it. The larger the surface area of the department's "front" (facing) and the brighter its colors, the more likely it is to be noticed and then bought, and the more brands and competition (national brands, private labels, etc.) "boost" this phenomenon.

The "inflationary" nature of the phenomenon is a cause for concern. One avenue to explore in the future is the widespread use of new "smart" labels, such as RFID (Radio Frequency Identification) labels. Combined with smartphone applications, they could contribute to this communication without overloading packaging, just as the digitalization of points of sale could lighten the merchandising aspects of packaging.

13.4. Service functions

The proliferation of packaging is also due to the fact that, in addition to technical and communication functions, packaging is becoming a source of innovation, not only in terms of content, but also in terms of the services provided by the container, and, more broadly, in terms of "packaging":

– "dispensing" and/or "dosing" pump systems for liquid soaps, creams, detergents, sauces and condiments;

– propellant gas aerosol for sprays (deodorants, water, perfumes, lacquers, paints, lubricants, etc.);

– applicator/doser systems (cosmetics, detergents);

– heating trays, individual briquettes, straws, cutlery, "plate boxes", glass cans and, in fast food, "snaking" products;

– individual pods for coffee machines;

– mixing systems for unstable products (injectable or drinkable solutions, cold welding, vulcanization).

13.5. Reflection points

The question is whether these products and the services they provide should be abandoned for ecological reasons. This raises a fundamental question for marketing and our industrial societies: "how far does the notion of 'need' go, in relation to a service provided, and when does it become a 'gadget'?"

Without seeking to answer this "existential" question, two axes of work are possible on these subjects:

1) reconsider these products according to their negative impact on the environment, or even the danger they represent to the planet, and prune, or even remove, them accordingly. Ban certain packaging, as already done for propellant gas aerosols containing CFCs, for example, or plastic straws;

2) make the fight against waste and the promotion of a circular economy a research axis for innovation in packaging-service, contributing to these new challenges: research on materials, manufacturing processes, user machines (in the case of pods), the establishment of PWR sectors in these areas (e.g. for fast food).

These innovations, while providing ever more services, create value and jobs.

13.6. Conclusion

As we have seen, decisions in terms of packaging design are complex to make, as the direct and indirect impacts on the environment can be numerous and even counterproductive. The options taken lead to ecological benefits as well as to certain disadvantages. By the end of this chapter, we can appreciate that public authorities are concerned about this, providing a more precise regulatory framework, in particular, for the organization of EPR sectors in which conflicting interests may arise. It also shows that an overall ecological balance sheet is needed to support packaging decisions. These decisions are no longer the responsibility of a Product Manager or operational marketing teams, but of the company's own management.

What to recommend?

– the creation of a multidisciplinary committee, in charge of studying packaging choices, including managers in purchasing, production, marketing-sales, supply chain, category management;

– this committee reporting to the sustainable development department, if it exists, or to the general management;

– membership in or the creation of "eco-responsible packaging" labels to reward the efforts of the most deserving companies;

Finally, three principles of the Sustainable Development approach seem particularly well-suited to the challenges related to packaging:

– the principle of transparency: companies throughout the supply chain – packaging producers, industrial users, logisticians, distributors – must have the courage to publish their figures, such as their CO_2 balance sheet (certified by independent bodies) in order to show, year after year, the progress made;

– the principle of continuous improvement: once an objective has been achieved, set a new and more ambitious one, for example, for the weight of the material used in a package;

– the principle of ecologically-responsible management: the environmental concern must prevail in managers' decisions.

The packaging and packaging policy is fully in line with the circular economy approach, first of all, through the emblematic principle of the deposit, to be updated and, far beyond that, through the search for better recyclable packaging and the implementation of more efficient recycling channels. Finally, this subject highlights the interaction and the powerful link between the evolution of production (design, development and more responsible use of packaging) and the evolution of consumption (vigilant, even militant attitude towards packaging, active participation in deposit and recycling systems). This link is even the main key factor for the success of this sector.

13.7. References

Fulconis, F. and Philipp, B. (2018). La question des emballages d'expédition dans l'E-commerce : une affaire de logistique et de marketing. *Projectique*, 3(21), pp. 23–41.

Geyer, R., Jambeck, J.R. and Lavender Law, K. (2017). Production, use and fate of all plastics ever made. *Science Advances*, 3(7).

Herbreteau, V. and Jacquot, M. (2017). Choix du couple emballage produit et role du transfert de sensation. *Industries agroalimentaires et agricoles,* January–February, pp. 30–32.

Pinet, B. (2004). Développement durable et marketing de l'emballage. *Revue Française du Marketing*, 200, pp. 27–38.

14

The Circular Economy and Toy Sector

Environmental issues such as the loss of biodiversity and climate change, as well as social inequalities, have been pointed out since the 1970s (Meadows *et al.* 1972). The circular economy, whose institutional foundations date back to the early 2000s, seems to be a path through which we could rethink our production and consumption patterns towards sustainable development.

However, today, there is not any official/controversial-free definition of the circular economy. For the *Institut National de l'Économie Circulaire*, the aim is:

> to move from an impact reduction model to a value creation model that is positive on a social, economic and environmental level[1]. [...] Its ultimate objective is to achieve the decoupling of economic growth from the depletion of natural resources through the creation of innovative products, services, business models and public policies.

In clearer terms, this means rethinking our production and consumption patterns. Murray *et al.* (2015) specify that in this model, "planning, procurement, purchasing, production and reprocessing are designed and managed, both as processes and outcomes, to maximize ecosystem functioning and human well-being." Therefore, circularity is based on the lifecycle (see Chapter 1 of this book) while involving a territorial and sectoral reading of the subject.

Despite many negative impacts, the toy sector has received little attention from the environment's and the circular economy's literatures. The very few existing analyses focus either on product safety, in which case legislation takes over, or on global value chains with their social abuses reported by NGOs and shared on social

---

Chapter written by Karen DELCHET-COCHET and Coralie DAMAY.

1 https://institut-economie-circulaire.fr/economie-circulaire/.

networks. The issue of waste is getting addressed, as evidenced by extended producer responsibility (EPR). EPR in the toy sector is drafted in the law on the waste and circular economy of fall 2019. But this does not necessarily involve a global approach. So, what really is the situation regarding the circular economy in this sector?

Moreover, toys are not goods like others. What do these very first objects we give our children show? And how can this impact the thinking in regard to the adoption of the circular economy in this sector?

14.1. The toy sector between impact and innovation

14.1.1. *The main characteristics of the toy sector*

A highly internationalized and highly seasonal sector

The games and toys sector[2] is a significant market. In 2018, it represented €3.4 billion in revenue in France[3]. Globally, seven groups produce 75% of the toys to be found on shelves: Mattel, Hasbro, Playmobil, V-Tech, Lego, Ravensburger and Bandai. This increasingly internationalized sector has experienced many relocations over the past 20 years. In France, however, we are witnessing a revitalization of the sector, supported by historical players as well as new brands whose position is based on sustainable development. Since 2014, French companies have been working together in the Association of French Toy Creators-manufacturers (*Association des créateurs-fabricants de jouets français* – ACFJF) in order to promote a geographical approach to production via the "Made in France" stamp. New players are positioning themselves with the principles of local and/or environment-friendly production. They are therefore partly involved in a circular economic perspective that involves "rethinking (the) models and production processes, from design to consumption, while integrating the need to recycle the components of products and even refurbish them" (Hoballah 2016). So, what is the situation, more generally, in this sector?

A normative and legislative framework under construction

The legislation on toys mainly focuses on the safety and toxicity of toys. As the website of the French Ministry of Economy and Finance points out, toys are regulated by the decree of 24 February 2010[4]. This refers to "essential safety requirements" and "CE marking", as well as the related conformity assessment. However, nothing is said about the social or environmental impact of the product.

---

2 In this work, we are interested in the games and toys sector, except video games.

3 https://www.fjp.fr/.

4 https://www.entreprises.gouv.fr/politique-et-enjeux/qualite/securite/la-reglementation-applicable-aux-jouets.

In other sectors such as textiles, the French legislator has developed a legal provision on extended producer responsibility (EPR). This obligation for the producer to finance the management of the end-of-life of its products helps to limit waste. To explain what an EPR is, the French Ministry of Ecological Transition has chosen a counter-example relating to toys (see Figure 14.1)! Indeed, in November 2019, there was no EPR for this sector. But this is being addressed as part of the anti-waste law for a circular economy.

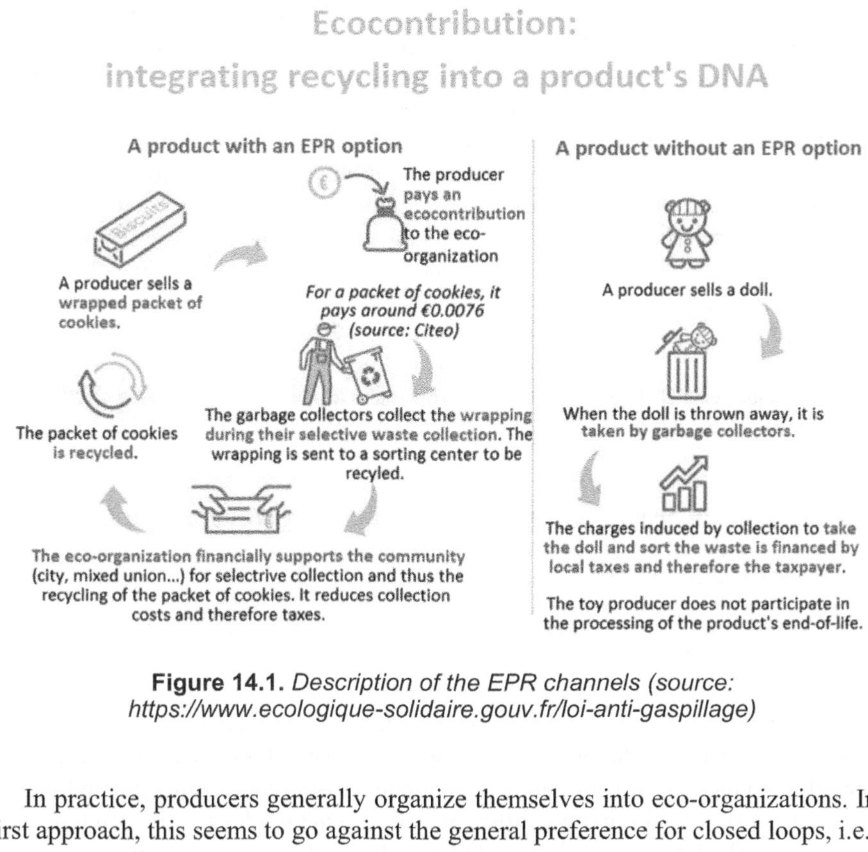

Figure 14.1. *Description of the EPR channels (source: https://www.ecologique-solidaire.gouv.fr/loi-anti-gaspillage)*

In practice, producers generally organize themselves into eco-organizations. In a first approach, this seems to go against the general preference for closed loops, i.e.:

> closed loop recycling and closed supply chains promoted by the circular economy in which a producer recycles and processes materials from his/her own end-of-life products for reuse in the same products in order to preserve and maintain the quality of materials for the longest period of time. (EMF 2013, cited by Kunz *et al.* 2018)

However, recent work by Geyer *et al.* (2016) on the environmental performances of open loops apparently provides more nuanced results. According to them, "closed-loop recycling is no better than open-loop recycling". For these authors, it is especially important to ensure that secondary raw materials, from recycling, actually substitute raw materials.

Even if the EPR process aims to encourage producers to design easy-to-recycle products (Kunz *et al.* 2018), one can only notice that companies are still very uncommitted to eco-design approaches as can be seen in other processes under EPR.

As for voluntary approaches, there is currently only one ecolabel for toys, cited by ADEME[5]: The Nordic Ecolabel. An ecolabel guarantees an environmentally friendly design for all the stages of the product's life. The Nordic Ecolabel for toys[6] more specifically bans named chemical substances, limits formaldehyde levels, guarantees the absence of genetically-modified trees in the wood used to make the product, and so on. It is therefore a guarantee of better environmental performance. Other environmental labels do exist but they are more specific. They will ensure that environmental criteria are taken into account in the production of the main raw materials, such as FSC or PEFC for sustainably-generated wood or Oeko-tex for organic fabrics. Finally, labels specifying the origin of the products can display quality claims such as "*Origine France Garantie*" (French origin guaranteed).

The possibilities for environmental labeling are therefore rather limited. What about the practices at each stage?

14.1.2. *The circular economy in the toy sector: a lifecycle perspective*

The importance of design

As Kunz *et al.* (2018) point out, "the majority of products are not currently designed to be easily reusable or reusable after recycling and, in many cases, require special treatment to eliminate hazardous materials at the end of their life." Toys, in their vast majority, are no exception to the rule. Design requires, from a product lifecycle perspective, integration of the environmental impacts at all stages: raw materials, production, distribution, use including reuse, repurposing and end-of-life. Each of these stages will involve different actors, generally located in different areas or countries, which makes thinking in terms of lifecycle even more complex.

But this does not mean that designers should be relieved of their responsibilities. Indeed, the environmental performance of some later stages may be directly dependent on the initial design. Let us therefore detail the characteristics of each step.

5 https://www.ademe.fr/labels-environnementaux.

6 http://www.nordic-ecolabel.org/product-groups/group/?productGroupCode=095.

The decisive choice of raw materials

The choice of raw materials is crucial. Today, toys are mainly made of plastic. Historically, plastic has been used in the toy sector since the 1960s. It was regarded as a source of innovation, as recalled by Barbe and Lioger (1999) in their analysis of companies in the Jura[7]. But the recent awareness of environmental issues is now pointing to this raw material made from oil, neither sustainable in terms of supply nor in terms of recyclability. There are, of course, different alternatives: natural materials such as rubber wood used by Plan Toys to make dolls houses, or bioplastic, made from sugar cane, for Lego bricks in 2018.

In addition, the toy is usually sold in a box. The design of the toy is therefore not without including the packaging. It is probably on this specific point that companies have made the biggest effort. However, despite their investments, much remains to be done, and this does not solve the problems posed by the lack of global thinking on the product.

Production, including logistics and transport

The production stage consists of the actual manufacture of the toys. This sector is not immune to the effects of globalization. Production is often outsourced, in countries with low labor costs and less stringent environmental legislation. Several companies, such as Mattel and Lego, have been identified by NGOs on this subject[8]. This step also involves rethinking manufacturing processes, so that they are the most economical in terms of non-renewable resources in particular.

For larger toy companies, production sites are usually far from the selling countries. However, Lego with its three factories located in Europe, Central America, and Asia is an exception to the above. The global flows of goods require sales in stores all over the world, with very strong seasonality. The impact in terms of transport is significant.

Distribution, towards a responsibility in terms of choice and message

Four categories of operators, in other words distributors, will be divided into this crucial stage: specialists in traditional toys (representing 40% of turnover), hypermarkets and supermarkets (32%), "pure players" in online sales (19%) and others (bazaars, sales outlets, department stores, traditional mail order companies representing 9%). Distributors are responsible for the choice of products distributed and the related communication, whether or not they raise consumer awareness of

7 Region of eastern France specializing in toy production.

8 https://www.novethic.fr/actualite/social/conditions-de-travail/isr-rse/mattel-en-chine-des-violations-du-code-du-travail-denoncees-depuis-15-ans-141622.html.

responsible consumption practices. The comments by Cho *et al.* (2018), who specify that "encouraging sustainable decisions and practices at the individual level is one of the necessary preconditions", fully underline the role of distributors as relays of information to the consumer. However, it must be noted that few have integrated these issues into their product choices and therefore into the related communication, outside specialized stores such as Nature et Découvertes or Bonhomme de Bois.

Use, between shelf life and new uses

The use stage involves two main complementary topics: the shelf life of the product before it is discarded and the duration of use, which may vary from one child to another and from one product to another.

The short lifespan of products is often due to the design itself. The role of toy manufacturers and designers is key to this. Toys may have been designed voluntarily with a short lifespan due to the poor quality of some components, for example. What about their reparability? The existence of spare parts to complete a game? Few manufacturers offer to repair their products like Corolle and its dolls' clinic. But this possibility is little known to the general public, complicated to implement and imposes conditions (such as an almost perfect condition of the doll) to be feasible. The short lifespan can also be related to trends, especially in the context of licensing. This seems to impose another form of programmed product obsolescence.

This subject would not be fully addressed without consideration of the short duration of the use of some toys in the life of the growing child. The role of toy libraries or dedicated structures such as the Playmobil Fun Park, which make it possible to use and/or rent toys, can be an interesting alternative, in terms of the functional economy.

When they no longer use the toy, consumers may show different types of habits:

– The toy is stored in order to pass it on, often within the same family, to a younger person. This choice makes it possible to keep the memories attached to the object by a new use instead of letting it gather dust.

– The toy is given to a known third party or charity to be used by someone else and here again find another use. In this case, it can be repaired and refurbished for sale in second-hand shops such as Rejoué (see Box 14.1) or Emmaüs, for reuse.

– The toy is sold in order to recover part of its cost and/or for educational purposes during car boot sales.

There are therefore many reuse practices in this sector.

The French association Rejoué is a specialist in the reuse of toys (see Figure 14.2). It collects toys, repairs them, cleans them and sells them in shops. In 2018, the association recovered 42,000 tons of toys and recovered two-thirds of them, thereby increasing product lifetime[9]. The cleaning of toys is done with ecological and biological products.

This approach helps to consider short loops for reuse.

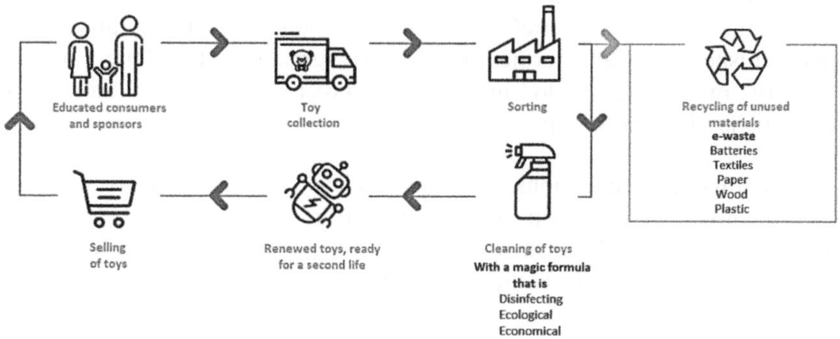

Figure 14.2. *Example of the integrated approach of REMPLOI by Rejoué, (source: https://rejoue.asso.fr/nos-missions-une-activite-eco-responsable/). For a color version of this figure, see iste.co.uk/delchet/circular.zip*

This company is also working on a positive approach to its social impact by employing people who are being reintegrated. In 2018, it was awarded the Association trophy by the *Institut National de l'Economie Circulaire.*

Box 14.1. *Rejoué*

– Finally, the toy can be thrown away when it is broken or when this solution is considered the easiest way to get rid of an object that is now considered useless or even cumbersome. Nothing is said about the end of its life. Most of it will be buried or incinerated, in other words, not recycled. Let us analyze this subject.

The end of product life and the limits of recycling

In recent years, the question of toy recycling has been raised by both the media and parents, who do not know what to do with the large amount of objects that are no longer suitable or even broken. According to the government, "75,000 tons of

9 https://france3-regions.francetvinfo.fr/paris-ile-de-france/jouets-se-recyclent-pere-noel-adepte-economie-circulaire-1387633.html.

toys end up in the garbage"[10]. Faced with this observation, as mentioned at the beginning of this chapter, the French government plans to carry out a toy EPR in order to encourage manufacturers in the sector to make a commitment to the end of the product's life.

Some manufacturers are beginning to work on the end of life of toys. For example, in April 2019, the giant Hasbro announced a partnership with the global waste treatment company TerraCycle. The Hasbro toys collected by this program are sorted, crushed and transformed into plastic granules. The latter are used as secondary raw material to manufacture urban benches or storage boxes, characteristic of an open loop.

While it has good intentions, recycling remains very weak for two reasons. First, recycling should only happen as a last resort, as specified in the law on energy transition for sustainable growth of August 18, 2015. Therefore, priority must be given to the reuse of products, then to their repurposing and then to their recycling. Second, collecting for recycling does not guarantee that the product is recyclable or even that it will actually be recycled. Toys are often made of several different materials or plastic, which makes the task even harder[11]. The importance of design is once against highlighted. Thus, Monopoly, G.I. Joe, My Little Pony and Transformers are now designed and manufactured with only one type of plastic, PET, which makes them easier to recycle[12].

Nevertheless, some toy manufacturers have a very integrated and committed approach, such as Bioviva[13]. This French company designs games, produced in a responsible way, in other words with the lowest possible environmental impact and a positive social impact. All steps are carefully looked into. But this company goes further by offering "caring games" that raise awareness of respect for the planet. This allows us to highlight another criterion: the purpose of the toy, which is not just an object as any other.

10 https://www.actu-environnement.com/ae/news/economie-circulaire-concertation-nouvelles-rep-jouets-velo-31576.php4.

11 https://www.liberation.fr/france/2019/07/13/comment-recycler-les-jouets-de-vos-enfants_1739209.

12 https://www.liberation.fr/france/2019/07/13/comment-recycler-les-jouets-de-vos-enfants_1739209.

13 https://www.bioviva.com/fr/.

14.2. A toy is more than "just a product"

Playing is the hallmark of contemporary childhood (Ariès 1960, cited by Gaussot 2001), whether this activity is perceived as frivolous (recreation or reward) or as a serious educational, pedagogical activity that contributes to the child's development. Since the beginning of the 20th Century, professionals working with children have agreed on this second perception. Playing is central for the child (Brougère 2003) and is often done thanks to a commercialized object: the toy. Childhood is perceived as a time of learning during which play and toys are often assimilated.

The toy has many facets. First of all, the toy as a means for playing contributes to the child's development both psychomotorally and in terms of identity building, socialization and even in school. Thus, the toy is an educational object, not only the basis for the personal construction of the youngest children, their skills and abilities, but also the basis for their understanding of their social environment (justice, cooperation, respect of rules and social relations according to age, gender, etc.). Market figures – the traditional toy market represented just under 3.5 billion euros in France in 2018[14], or an annual consumption of 275 euros per year per child aged 0–11[15] – seems to testify parents' support for this observation. Toy professionals go even further because for them: playing is life[16].

14.2.1. *Toys, the company's flagship products*

In a socio-economic dimension, the toy reflects the values and foundations of our society. As such, board games can be seen as a vehicle for children to learn the rules of our consumer societies: sell, buy, anticipate the reaction of other players, and so on.

From a socio-cultural perspective, it is recognized that children are sensitive to cultural gender stereotypes and that toys contribute to their sexual differentiation (Brougère 2003). According to this conception, the toy is designed according to the role, or future role, assigned to boys vs. girls. Toys for boys can therefore be seen as a catalyst for their aggressiveness and encourage them to compete. For girls, however, their maternal fiber and domestic chores are exacerbated. Even if debates on the gender "signs" provided by toys have existed for several years, the recurring polemics around Christmas catalogues clearly illustrate that the issue remains topical and that expectations towards producers remain high (e.g. Ezan and Ulrich 2016).

14 www.fjp.fr.

15 www.fjp.fr.

16 www.fjp.fr.

Finally, from an educational perspective, the studies highlight the various roles of toys in the family education project. In middle and upper classes, toys are learning tools, while in lower classes it is a reward or sanction for academic achievements/failures (Vincent 2000). Therefore, the importance of toys is different according to families.

14.2.2. *A brief historical return*

Understanding the place of toys cannot be done without drawing a parallel with the place given to children in our society. At the end of the war, educational and psychological importance was given to children. At the same time, birth control and the increasing number of working women led to children being considered as child-kings in the 1970s. In a period of claiming the right to happiness and personal fulfillment, in opposition to the values of authority and obedience that previously prevailed, the desired child was a subject in their own right. Families wanted to make their child happy and it became the focus of their attention. During the *Trente Glorieuses*, the 30 years following the end of World War II, the place given to the child, and the increasing guilt of some parents whose time available for their families was decreasing, explains the use and development of toys. Toys are used to make up for the short time granted to children using entertaining products made to contribute to their development. It gradually became a key element of their development.

Toy manufacturers ride on the waves of this parental expectation – the desire to offer the best to their child and to help them become a successful grown-up – and offer a range of accessories to create skill-learning situations (Damay and Llorca 2012):

– learning to read and count are offered by many toys for young children from magnetic letters, to wooden abacuses and mini computers;

– learning to manage your pocket money and the value of things is possible through merchant toys as well as many board games, the most famous of which, Monopoly, allows you to buy properties or hotels;

– learning to choose, to make decisions, to understand your limits as proposed by strategy games as well as all the toys to collect and exchange in the playground;

– learning to express yourself in public becomes fun with many games that allow you to communicate through drawings or without saying a list of forbidden words;

– learning to behave in society is *encouraged* by *most* toys and games which, when they involve several players, require agreement on the rules and the ability to behave with others.

14.2.3. *The role of the toy in educating the future consumer?*

The child's change of status in our society, now considered as an individual with their own rights, contributes to giving a child multiple roles. They are a buyer thanks to having their own money (pocket money, tooth fairy, etc.). They are also a direct prescriber when they make a request, or an indirect prescriber when they refuse to consume in order to show their disagreement (e.g. a refusal to eat yogurt). Finally, they are a future consumer insofar as their learning (skills and knowledge) during childhood will largely be mobilized in adulthood (McNeal 1992; Belk *et al.* 1984). This 3-in-1 role, that is now recognized, explains the interest granted to children by different stakeholders such as researchers (in human sciences as well as in management sciences), public authorities and companies. For companies, having a precise knowledge of this population allows them to offer goods matching children's real needs (psychological, nutritional, etc.) and expectations. However, the role of marketing must be questioned. Indeed, the over-mediatization of toys, the importance of licenses, the communications of certain professionals and so on can be perceived by adults as strong injunctions and make them feel guilty for not buying a specific toy for their child. The parent or adult therefore feel they are jeopardizing their child's development, keeping them from certain opportunities.

But, as early as the 1970s, Kotler and Zatlman (1971) mentioned a counterpart to commercial marketing: social marketing. Based on the same techniques, social marketing aims to voluntarily change consumer's behaviors to improve the well-being of the target and society. In the same period of time, a research stream on a conserver society that would "maintain or improve one's standard of living while consuming less or differently" (LeGall 2002, p6) emerged and developed, before declining in the 1980s because of the importance given to ownership (LeGall 2002). Nowadays, social marketing is mobilized as part of prevention of obesity, smoking and so on among children and teenagers. Given the challenges of sustainable development and the growing interest in sustainable consumption, why not think about its use in the toy sector? Social marketing could help reduce overconsumption habits and promote a more restrained way of consuming that is essential to protect our planet.

14.2.4. *Toys and environmental education*

Historically, our society was led by consumerism. However, resistance is appearing. New and more responsible ways of consuming are developing. Shared between performance injunctions for their children and the sacralization of childhood values – such as carelessness and spontaneity – parents are starting to reject a consumption that is considered excessive. For example, they will develop their children's critical sense, make them aware of the different sales techniques, make them think about commercial objectives of advertising and so on. More specifically, parental resistance exists in the toy sector (Nabec 2013). If the toy is a hedonic and involving product because it carries educational values and is a path to socialization, some parents nevertheless wish to teach their children to consume less, consume differently or not at all (Nabec 2013). For these parents, the relationship with toys is at the heart of their educational method. This observation then encourages professionals of the sector to modify their offer and reposition it. They should do this by asking themselves: to what extent could toys become a support for learning sustainable consumption?

Moreover, while work on children and the environment is relatively recent and still very limited (Francis and Davis 2015), it is important to stress that early results tend to show interest in environmental issues and a positive attitude towards the environment (Evans *et al.* 2007; Grønhøj and Thøgersen 2012, cited by Parguel and Charry 2019). Children show interest in the climate, pollution, animal survival, deforestation, well-being and so on, whatever their cultural or economic backgrounds (Francis and Davis 2015). At school age, they are motivated to protect the environment and pass onto their parents the lessons learned in school (e.g. Zelezny 1999; Vaughan *et al.* 2003). More specifically, children are involved and aware of the phenomena of sorting, reflecting a certain know-how (Schill and Fosse-Gomez 2014). Therefore, other professionals in children's products have already taken into account their sensitivity and skills in following advice, as shown, for example, by the newspaper Astrapi of November 15, 2019, which offers a press kit entitled "Save the Earth!" for children aged 7–11 years with challenges to be met as a family.

However, there is little research on how to educate and encourage pro-environmental behaviors in children aged 7–12 years (Parguel and Charry 2019). In the toy sector studied in this research, two means appear. First, to propose toys aiming to raise awareness of environmental issues or citizenship in the broad sense. Second, to offer toys that are exemplary in terms of design, reflecting a circular economy approach or at least the company's own eco-design. In the first case, it is the purpose of the toy that has been questioned. What will the toy be used for? What is the objective of the game? Many toys are already designed with an educational vocation (learning to count, tying shoelaces, mapping the world, etc.), but few are developed with a goal to raise collaborative or ecological awareness.

14.2.5. *The toy, a role to be redefined*

Given the challenges society faces, it seems necessary to invite stakeholders to consider the role given to the toy. Indeed, if playing is intrinsically linked to the child, should the toy, and more precisely all toys, automatically be included as an essential element for a child's development? If the answer were a straightforward yes, it would mean denying the development of the poorest children whose access to toys is sometimes more than limited. To this tricky question, a simple answer cannot be satisfactory and we invite all stakeholders to question the utility of the toy and not (automatically) take it for granted.

The toy as only a toy does not seem necessary. But the toy that encourages development may have its uses. It is then a catalyst that helps children to mime, to reproduce the gestures they daily witness. The question seems to be about the level of similarity between the toy and the object of the adult's world. Imitation is an important element of child development, but should it be accepted as obvious that this imitation involves a concrete object? Is it necessary to have a kitchen set with a miniature food processor of the same brand as the parents'?

Harder still is the questioning of the toy as a symbolic object. Toys are often a strong marker of social status and a tool for the child to join a peer group. Nevertheless, the trend of licensing and trendiness need to be questioned. The socialization of children with others cannot result solely from the possession of one toy or another, as this could lead to children being excluded from playgroups. Therefore, limiting the effects of trends and designing toys with a purpose other than short-term sales should become an issue for all stakeholders (parents but also industrialists, distributors and public authorities).

Toys provide us with information about our society. If we only provide younger generations with objects that mirror consumption, it will certainly be more complicated to make them aware of the increasing scarcity of raw materials, pollution and restraint. We note that very few collaborative toys exist. The most common rule is: win and win over other players! This observation leads us to ask toy manufacturers to develop an offer that allows them to play truly together and not against each other.

Another important aspect is that creative toys can be made with toxic and polluting substances, even though the usefulness of these toys as facilitators of imagination and creativity is often well grounded. It is important to note however that creativity and cooperation are values sought (Taddei 2018) to find innovative solutions for the future. As for construction toys, which promote imagination and spatial representation, they are often pointed out for their perceived sexist nature.

The question of the usefulness of toys leads us to make proposals to rethink this very special universe.

14.3. To conclude: proposals for rethinking our relationship with toys

Following these observations, we propose here some ways to rethink our relationship with toys and to place this sector in a circular economy perspective:

– One of the key points is to have a product lifecycle approach. This reduces environmental impacts for all stages of the toy's lifecycle. The toy is not only designed as recyclable. It may be made of compostable materials, will be repairable, reusable and so on. This design approach also entails, as far as possible, integrating geographical proximity into the reflection process. This subject is fundamental both in terms of reducing energy/climate impacts and in order to be able to consider flow management between companies in the same territory. it is referred to as territorial ecology.

– This study highlights the importance of a brainwork on the purpose of toys. While it is necessary to conceive toys differently from a technical point of view, the importance of thinking about their usefulness and the values they convey should not be overlooked. Thus, we believe it is necessary to develop games that will enable future eco-citizens to be trained instead of consumers and to value collaboration and creativity. More generally, despite the many questions raised by the circular economy, the question of the usefulness of the product does not always seem to be sufficiently emphasized. At the time the good is purchased, the individual should ask themselves: do I really need this object? When purchasing a toy, as with any product, we invite people to ask themselves these questions: what will be the use of this toy? What child or teenager's needs will this toy meet?

– Toys designed in this way must be supported, valued and chosen by all stakeholders: manufacturers, distributors (who select certain products), buyers in the broad sense, individuals or organizations (childcare facilities – nurseries, kindergartens, leisure centers –, schools, etc.), public authorities, etc. It seems essential for the successful integration of the toy sector into the circular economy (more than the integration of the circular economy into the toy sector) that this achievement be driven by all and not a single stakeholder.

– All these practices also need to be known, both by adults and children. Distributors have a fundamental role in promoting eco-friendly products and informing adult buyers, who have less and less time to spend on purchases. For example, advice at the time of purchase, clear information (available in catalogues, advertisements, etc.) must be provided.

– The circular economy also requires a redesign of the business model (Geissdoerfer 2018). Developing new business models based on, for example, the economy of functionality, i.e. the rental of products, could be considered. The creation of new toy libraries could also participate in this circular economy via the sharing economy. Another avenue is the development of second-hand toy stores to increase the life of the product. These initiatives would make it possible to move towards restraint while being economically viable.

14.4. References

Ariès, P. (1960). *L'Enfant et la vie familiale sous l'Ancien Régime*. Plon, Paris.

Barbe, N. and Lioger, R. (1999). Du jouet en bois au jouet en plastique. In *Innovation et culture technique dans l'arc jurassien*, Christian Bromberger and Denis Chevallier (eds). Editions La maison de l'homme, Paris, pp. 43–58.

Belk, R., Mayer, R. and Driscoll, A. (eds) (1984). Children's recognition of consumption symbolism in children's products. *Journal of Consumer Research*, 10, no. 4, pp. 386–397.

Brougère, G. (2003). *Jouet et compagnie*. Stock, Paris.

Cho, Y.-N., Soster, R.L. and Burton, S. (eds) (2018). Enhancing environmentally conscious consumption through standardized sustainability information. *The Journal of Consumer Affairs*, 52, no. 2, pp. 393–414.

Damay, C. and Llorca, A. (2012). Les enfants dans l'espace commercial. In *Kids Marketing*, 2nd edition, Brée, J. (ed.). EMS Management & Société, Cormelles-le-Royal, pp. 413–448.

Ellen MarcArthur Foundation (2013). Toward a circular economy [Online]. Available at: https://www.ellenmacarthurfoundation.org/assets/downloads/publications/Ellen-MacArthur-Foundation-Towards-the-Circular-Economy-vol.1.pdf [accessed 10 October 2019].

Evans, G.W., Brauchle, G., Haq, A., Stecker, R., Wong, K. and Shapiro, E. (eds) (2007). Young children's environmental attitudes and behaviors. *Environment and Behavior*, 39, no. 5, pp. 635–658.

Ezan, P. and Ulrich, I. (2016). Bouleverser les codes dans les catalogues de jouets : réactions des enfants et légitimité de la démarche. *Décisions marketing*, 82, pp. 53–72.

Francis, J.E. and Davis, T. (2015). Adolescents' sustainability concerns and reasons for not consuming sustainably. *International Journal of Consumer Studies*, 39, pp. 43–50.

Gaussot, L. (2001). *Le jeu de l'enfant et la construction sociale de la réalité, Le Carnet PSY*. Editions Cazaubon, 2, no. 62, pp. 22–29.

Geissdoerfer, M., Morioka, S.N., Monteiro de Carvalho, M. and Evans, S. (2018). Business models and supply chain for the circular economy. *Journal of Cleaner Production*, 190, pp. 712–721.

Geyer, R., Kuczenski, B., Zink, T. and Henderson, A. (eds) (2016). Common misconceptions about recycling. *Journal of Industrial Ecology*, 20, no. 5, pp. 1010–1017.

Grønhøj, A. and Thøgersen, J. (2012). Action speaks louder than words: The effect of personal attitudes and family norms on adolescents' pro-environmental behaviour. *Journal of Economic Psychology*, 33, no. 1, pp. 292–302.

Hoballah, A. (2016). Préface. In *L'économie circulaire, une transition incontournable*, Sauvé, S., Normandin, D. and McDonald, M. (eds). Presses Universitaires de Montréal, Montreal.

Kotler, P. and Zaltman G. (1971). Social marketing: An approach to planned social change. *Journal of Marketing*, 35, no. 3, pp. 3–12.

Kunz, N., Mayers, K. and Van Wassenhove, L.N. (eds) (2018). Stakeholder views on extended producer responsibility and the circular economy. *California Management Review*, 60, no. 3, pp. 45–70.

Le Gall, M. (2002). De la préoccupation pour l'environnement à la consommation durable. In *2ème Congrès sur les Tendances du Marketing en Europe*, 25–26 January.

Meadows, D., Meadows, D., Randers, J. and Behrens, W.W. (eds) (1972). *Halte à la croissance?* Fayard, Paris.

McNeal, J.U. (1992*). Kids as Consumers*. Lexington Books, New York.

Moore, E.S. and Lutz, R. (1988). Intergenerational influences in the formation of customers' attitudes and beliefs about the marketplace/mother and daughters. *Advances in Consumer Research*, 15, pp. 461–467.

Moore, E.S., Wilkie, W.L. and Lutz, R. (eds) (2002). Passing the torch: Intergenerational influences as source of brand equity. *Journal of Marketing*, 66, pp. 17–37.

Moschis, G.P. and Churchill, G.A. (1978). Consumer socialization: A theoretical and empirical analysis. *Journal of Marketing Research*, 15, no. 4, pp. 599–609.

Murray, A., Skene, K. and Haynes, K. (2015). The circular economy: An interdisciplinary exploration of the concept and application in a global context. *Journal of Business Ethics*, 140, pp. 369–380.

Nabec, L. (2013). Les formes de résistance parentale à la consommation enfantine et au kids marketing. *Management & Avenir*, 2, no. 60, pp. 157–175.

Parguel, B. and Charry, K. (2019). Educating children to environmental behaviours with nudges: The effectiveness of social labelling and moderating role of age. *Environmental Education Research*, pp. 1–15.

Riesman, D. and Roseborough, H. (1955). Careers and consumer behavior. In *Consumer Behavior Vol. II: The Life Cycle and Consumer Behavior*, Clark, L.H. (ed.). Lincoln B. York University Press, pp. 1–18.

Roedder John, D. (1999). Consumer socialization of children: A retrospective look at twenty-five years of research. *Journal of Consumer Research*, 26, no. 3, pp. 183–213.

Schill, M. and Fosse-Gomez, M.H. (2014). La place de l'enfant dans le comportement écologique de la famille : une approche par le mime. In *13ème Journées Normandes de Recherches sur la consommation : Société et Consommation*, Rouen.

Taddei, F. (2018). *Apprendre au XXI*ème *siècle*. Calmann Levy, Paris.

Ulrich, I. and Ezan, P. (2016). Boys and dolls, girls and cars. *International Journal of Retail & Distribution Management*, 44, no. 10, pp. 1047–1063.

Vaughan, C., Gack, J., Solorazano, H. and Ray, R. (eds) (2003). The effect of environmental education on schoolchildren, their parents, and community members: A study of intergenerational and intercommunity learning. *The Journal of Environmental Education*, 34, no. 3, pp. 12–21.

Vincent, S. (2000). Le jouet au cœur des stratégies familiales d'éducation. *Sociétés contemporaines*, 4, no. 40, pp. 165–182.

Ward, S., Wackman, D.B. and Wartella, E. (eds) (1977). *How Children Learn to Buy*. Sage, Beverly Hills.

Zelezny, L.C. (1999). Educational interventions that improve environmental behaviors: A meta-analysis. *The Journal of Environmental Education*, 31, no. 1, pp. 5–14.

Postface

From Inebriety to Sobriety: *Because Gê Is Worth It!*

The exciting thing about the circular economy is that it simply forces us to reinvent the world, the relationships between humans, the relationship between humanity and the rest of life, in short our relationship to each other, whatever it is: human, animal, plant or mineral.

It is a mistake to think that the current dominant model is the ideal balance that should not be changed. First of all, there is an ontological error because it is the intrinsic nature of human beings to adapt and evolve at the same time as their environment; second, there is an error facing future generations by letting them pay our environmental debts, which they already rightly blame us for.

A shared sense of urgency

We have no choice: the findings of the three international UN bodies, IPPC (Intergovernmental Panel on Climate Change), IRP (International Resource Panel) and IPBES (Intergovernmental Science-Policy Platform on Biodiversity and Ecosystem Services), are clear on the direction we are taking. Some people criticise these bodies for their supposed pessimism, or even for knowingly blackening the landscape to create concern, anguish and force political decision-makers to take radical measures.

It means forgetting that, as scientists, they only draw up observations and perspectives with models based on the knowledge of the moment. In 1972, the Club of Rome published the so-called Meadows report entitled *The Limits to Growth*,

---

Chapter written by Jean-Louis BERGEY

which also generated a lot of negative criticism, particularly in terms of methodology. Forty years later, Australian researchers[1] analyzed the evolution of the six indicators in the Meadows report, between 1970 and 2000, and demonstrated that they followed the trend predicted by the report.

So let us take seriously the IRP, which foresees a doubling of raw material consumption by 2060[2] and considers this to be an "unsustainable future in terms of both resource use and emissions, probably exceeding all possible measures of available resources [...]".

It also means concealing the fact that other organizations, including those who support the orthodox macroeconomic model of growth, share the findings. For example: the OECD on the economic and quality of life consequences of commodity consumption prospects[3]; or IMF President Kristalina Georgievia, who, at her press conference on taking office in 2019, stated that climate risk will now be included in economic analyses[4]; or the new President-elect of the European Commission, Mrs Ursula von der Leyen, who prioritizes a Green Deal for Europe; or the 40 or so large French or international companies grouped within EpE (*Entreprises pour l'Environnement*) which, in a report entitled Zen 2050[5], acknowledge the climate emergency.

The limit of a purely technological approach

Of course, most of the solutions advocated by these organizations are based on science and technology to help us reduce our consumption of fossil fuels, metals, materials, water, soil or biomass and, *in fine,* to make a relative decoupling between (positive) GDP trends and the consumption of natural resources. Others (companies and space agencies) go further, preparing mining projects on the Moon or Mars, after having started mining resources in the deep sea and polar regions.

And then what? For every resource reaches its limits if we are in the "ever more", regardless of the progress science makes to push it back, with a world

1 Turner, G. (2014). Is Global Collapse Imminent? MSSI research paper no. 4, University of Melbourne.

2 UNEP (2011) Decoupling natural resource use and environmental impacts from economic growth. United Nations Environment Programme.

3 OECD (2019). Business Models for Circular Economy: Opportunities and Challenges for Policy. Report, OECD Publishing, Paris.

4 *La Tribune*, 18/10/2019.

5 EpE (2019). ZEN 2050 imaginer et construire une France neutre en carbone. 20 May, France.

population of 11 billion inhabitants by 2050. This is what the American economist and philosopher Kenneth Boulding summed up well in the following maxim:

"Anyone who believes in indefinite growth in anything physical, on a physically finite planet, is either mad or an economist."[6]

Of course, we have made progress in all areas. But technology itself has its limits and is only worthwhile if it is used properly, which is not always the case, especially with the rebound effect. For example, in the automotive sector, progress in fuel consumption per kilometer has been reversed over time with the increase in mileage driven.

Deep changes needed to move towards sobriety

We must therefore go further, as the United Nations itself encourages us to do so by indicating in its report published in November 2019 on the gap between the needs and prospects for reducing GHG emissions: "Deep-rooted shifts in values, norms, consumer culture and world view are inescapably part of the great sustainability transformation".

The ultimate path of this change of culture, beyond all the actions to be undertaken in the field of the circular economy, is to enter the era of sobriety, consented to if not desired (and if possible, happy, as proposed by Pierre Rabhi[7]), but assumed. This in no way means "a return to the candle" but another way of conceiving our economy, meeting our needs, monitoring indicators, in short, a paradigm shift, which is obviously not an easy task. It is therefore necessary for the third of humanity that consumes too much to consume less, and not only less thanks to technology, but less, quite simply, thus allowing the other two-thirds who cannot satisfy their basic needs to do so, despite everything: the cake remains identical, but we share it better.

All this effectively reassesses the foundations on which our modern society has developed: What is the business model for companies producing less? What about the future of those working in manufacturing? What about salary? What about working hours? What about the trade-off between the collective and the individual? Incentive policy or coercive policy? What about the representation of happiness, completely dependent on the consumption of objects and of which we are reminded in every advertisement? And our relationship to money, to "wealth" which is only

6 United States Congress House (1973) Energy reorganization act of 1973: Hearings, Ninety-third Congress, first session, on H.R. 11510, p. 248.

7 French writer, farmer and environmentalist who promotes the ethics of "happy sobriety".

that of money among all others. We must also change our relationship to the object in that it allows us to satisfy wishes or desires, as well as to integrate ourselves into a group or on the contrary to distinguish ourselves from the mass.

We must change our relationship to time, both the one that drives us to ever more technology and more dependence on networks (electricity, transport, telephony), the slightest breakdown of which becomes a disaster, as well as the one of economic reflection and analysis, which is too short, preventing us from seeing the impacts in the long term.

It is necessary to change the measurement instrument, GDP obsessively supplanting all the others and carrying in itself this economic contradiction that it does not integrate the depreciation of natural capital, which would not be accepted for any company if it were its share capital. Changing the indicator would help us to change our perspective, except that changing the indicator is already having done most of the work.

But who should start this virtuous loop? Political leaders? Companies? Individuals? Actually all three!

To do this, we will have to overcome all our fears and take risks: the elected representative in the next election, the company on its annual result and the citizen in the comfort of their habits. However, taking risks does not mean losing every time. What satisfaction the political leader has for improving the lives of their fellow citizens, for the entrepreneur to have positioned themself as a leader in a new market and for the citizen to live in a more pleasant and healthy environment.

As we can see, we are only at the beginning of a large movement, inevitably difficult because it will be so profound and still paved with uncertainties, but necessary and even indispensable. This book illustrates, if necessary, the importance of a collective and multidisciplinary approach to understanding the issues related to the circular economy, in order to be part of this inevitable ecological transition and the importance of being vigilant to ensure that this circular economy is no less disabling and restricted.

The best for us is that we chart this new path together and from now on, even if we do not master all the parameters (but have we always mastered everything?) and that we accompany this change rather than undergo it and have to decide urgently, which is never the right solution.

In short, think about the future to better design it.

List of Authors

Franck AGGERI
CGS-i3 UMR
CNRS 9217
MINES ParisTech
France

Mérylle AUBRUN
AFNOR Group
La Plaine Saint-Denis
France

Vincent AUGISEAU
UniLaSalle
and
CitéSource
Rennes
France

Anne BENADY
AFNOR Group
La Plaine Saint-Denis
France

Jean-Louis BERGEY
ADEME
Angers
France

Dominique BONET FERNANDEZ
IPAG Business School
Paris
and
CRET-LOG
Aix-Marseille University
France

Bénédicte BOURCIER-BÉQUAERT
ESSCA School of Management
Aix-en-Provence
France

Christian BRODHAG
Ecole des Mines de Saint-Etienne
and
Pôle National Ecoconception
Saint-Etienne
and
Construction 21
Paris
France

François CABARET
Groupe ISC Paris
France

Coralie DAMAY
Groupe ISC Paris
France

Anne DE BÉTHENCOURT
Reset The World
and
C2C Community
Paris
France

Karen DELCHET-COCHET
Groupe ISC Paris
France

Karine FABRE
Senior Territorial Attaché
Paris
France

Valérie FERNANDES
CERIIM
La Rochelle Business School
CEREGE EA1722
France

François-Michel LAMBERT
French Parliament
Paris
and
Institut National de l'Économie
Circulaire
Paris
France

Catherine LEJEALLE
Groupe ISC Paris
France

Melodie MERENDA
AFNOR Group
La Plaine Saint-Denis
France

David MORIEZ
Groupe ISC Paris
France

Yvon PESQUEUX
HESAM University
and
CNAM
Paris
France

Alexis POKROVSKY
LIRSA (EA4603)
CNAM
Paris
France

Index

H, I, L

M, N, P

R, S, T

U, V, W

Other titles from

ISTE

in

Innovation, Entrepreneurship and Management

2020

ANDREOSSO-O'CALLAGHAN Bernadette, DZEVER Sam, JAUSSAUD Jacques, TAYLOR Richard
Sustainable Development and Energy Transition in Europe and Asia
(Innovation and Technology Set – Volume 9)

CORON Clotilde
Quantifying Human Resources: Uses and Analysis
(Technological Changes and Human Resources Set – Volume 2)

CORON Clotilde, GILBERT Patrick
Technological Change
(Technological Changes and Human Resources Set – Volume 1)

CERDIN Jean-Luc, PERETTI Jean-Marie
The Success of Apprenticeships: Views of Stakeholders on Training and Learning
(Human Resources Management Set – Volume 3)

DIDAY Edwin, GUAN Rong, SAPORTA Gilbert, WANG Huiwen,
Advances in Data Science
(Big Data, Artificial Intelligence and Data Analysis Set – Volume 4)

DOS SANTOS PAULINO Victor
Innovation Trends in the Space Industry
(Smart Innovation Set – Volume 25)

GUILHON Bernard
Venture Capital and the Financing of Innovation
(Innovation Between Risk and Reward Set – Volume 6)

MAKRIDES Andreas, KARAGRIGORIOU Alex, SKIADAS Christos H.
Data Analysis and Applications 3: Computational, Classification, Financial, Statistical and Stochastic Methods
(Big Data, Artificial Intelligence and Data Analysis Set – Volume 5)
Data Analysis and Applications 4: Financial Data Analysis and Methods
(Big Data, Artificial Intelligence and Data Analysis Set – Volume 6)

MASSOTTE Pierre, CORSI Patrick
Complex Decision-Making in Economy and Finance

MORLAT Clément
Sustainable Productive System: Eco-development versus Sustainable Development
(Smart Innovation Set – Volume 26)

SAULAIS Pierre, ERMINE Jean-Louis
Knowledge Management in Innovative Companies 2: Understanding and Deploying a KM Plan within a Learning Organization
(Smart Innovation Set – Volume 27)

2019

AMENDOLA Mario, GAFFARD Jean-Luc
Disorder and Public Concern Around Globalization

BARBAROUX Pierre
Disruptive Technology and Defence Innovation Ecosystems
(Innovation in Engineering and Technology Set – Volume 5)

DOU Henri, JUILLET Alain, CLERC Philippe
Strategic Intelligence for the Future 1: A New Strategic and Operational Approach
Strategic Intelligence for the Future 2: A New Information Function Approach

FRIKHA Azza
Measurement in Marketing: Operationalization of Latent Constructs

FRIMOUSSE Soufyane
Innovation and Agility in the Digital Age
(Human Resources Management Set – Volume 2)

GAY Claudine, SZOSTAK Bérangère L.
Innovation and Creativity in SMEs: Challenges, Evolutions and Prospects
(Smart Innovation Set – Volume 21)

GORIA Stéphane, HUMBERT Pierre, ROUSSEL Benoît
Information, Knowledge and Agile Creativity
(Smart Innovation Set – Volume 22)

HELLER David
Investment Decision-making Using Optional Models
(Economic Growth Set – Volume 2)

HELLER David, DE CHADIRAC Sylvain, HALAOUI Lana, JOUVET Camille
The Emergence of Start-ups
(Economic Growth Set – Volume 1)

HÉRAUD Jean-Alain, KERR Fiona, BURGER-HELMCHEN Thierry
Creative Management of Complex Systems
(Smart Innovation Set – Volume 19)

LATOUCHE Pascal
Open Innovation: Corporate Incubator
(Innovation and Technology Set – Volume 7)

LEHMANN Paul-Jacques
The Future of the Euro Currency

LEIGNEL Jean-Louis, MÉNAGER Emmanuel, YABLONSKY Serge
Sustainable Enterprise Performance: A Comprehensive Evaluation Method

LIÈVRE Pascal, AUBRY Monique, GAREL Gilles
Management of Extreme Situations: From Polar Expeditions to Exploration-Oriented Organizations

MILLOT Michel
Embarrassment of Product Choices 2: Towards a Society of Well-being

N'GOALA Gilles, PEZ-PÉRARD Virginie, PRIM-ALLAZ Isabelle
Augmented Customer Strategy: CRM in the Digital Age

NIKOLOVA Blagovesta
The RRI Challenge: Responsibilization in a State of Tension with Market Regulation
(Innovation and Responsibility Set – Volume 3)

PELLEGRIN-BOUCHER Estelle, ROY Pierre
Innovation in the Cultural and Creative Industries
(Innovation and Technology Set – Volume 8)

PRIOLON Joël
Financial Markets for Commodities

QUINIOU Matthieu
Blockchain: The Advent of Disintermediation

RAVIX Joël-Thomas, DESCHAMPS Marc
Innovation and Industrial Policies
(Innovation between Risk and Reward Set – Volume 5)

ROGER Alain, VINOT Didier
Skills Management: New Applications, New Questions
(Human Resources Management Set – Volume 1)

SAULAIS Pierre, ERMINE Jean-Louis
Knowledge Management in Innovative Companies 1: Understanding and Deploying a KM Plan within a Learning Organization
(Smart Innovation Set – Volume 23)

SERVAJEAN-HILST Romaric
Co-innovation Dynamics: The Management of Client-Supplier Interactions for Open Innovation
(Smart Innovation Set – Volume 20)

SKIADAS Christos H., BOZEMAN James R.
Data Analysis and Applications 1: Clustering and Regression, Modeling-estimating, Forecasting and Data Mining
(Big Data, Artificial Intelligence and Data Analysis Set – Volume 2)
Data Analysis and Applications 2: Utilization of Results in Europe and Other Topics
(Big Data, Artificial Intelligence and Data Analysis Set – Volume 3)

VIGEZZI Michel
World Industrialization: Shared Inventions, Competitive Innovations and Social Dynamics
(Smart Innovation Set – Volume 24)

2018

BURKHARDT Kirsten
Private Equity Firms: Their Role in the Formation of Strategic Alliances

CALLENS Stéphane
Creative Globalization
(Smart Innovation Set – Volume 16)

CASADELLA Vanessa
Innovation Systems in Emerging Economies: MINT – Mexico, Indonesia, Nigeria, Turkey
(Smart Innovation Set – Volume 18)

CHOUTEAU Marianne, FOREST Joëlle, NGUYEN Céline
Science, Technology and Innovation Culture
(Innovation in Engineering and Technology Set – Volume 3)

CORLOSQUET-HABART Marine, JANSSEN Jacques
Big Data for Insurance Companies
(Big Data, Artificial Intelligence and Data Analysis Set – Volume 1)

CROS Françoise
Innovation and Society
(Smart Innovation Set – Volume 15)

DEBREF Romain
Environmental Innovation and Ecodesign: Certainties and Controversies
(Smart Innovation Set – Volume 17)

DOMINGUEZ Noémie
SME Internationalization Strategies: Innovation to Conquer New Markets

ERMINE Jean-Louis
Knowledge Management: The Creative Loop
(Innovation and Technology Set – Volume 5)

GILBERT Patrick, BOBADILLA Natalia, GASTALDI Lise,
LE BOULAIRE Martine, LELEBINA Olga
Innovation, Research and Development Management

IBRAHIMI Mohammed
Mergers & Acquisitions: Theory, Strategy, Finance

LEMAÎTRE Denis
Training Engineers for Innovation

LÉVY Aldo, BEN BOUHENI Faten, AMMI Chantal
Financial Management: USGAAP and IFRS Standards
(Innovation and Technology Set – Volume 6)

MILLOT Michel
Embarrassment of Product Choices 1: How to Consume Differently

PANSERA Mario, OWEN Richard
Innovation and Development: The Politics at the Bottom of the Pyramid
(Innovation and Responsibility Set – Volume 2)

RICHEZ Yves
Corporate Talent Detection and Development

SACHETTI Philippe, ZUPPINGER Thibaud
New Technologies and Branding
(Innovation and Technology Set – Volume 4)

SAMIER Henri
Intuition, Creativity, Innovation

TEMPLE Ludovic, COMPAORÉ SAWADOGO Eveline M.F.W.
Innovation Processes in Agro-Ecological Transitions in Developing Countries
(Innovation in Engineering and Technology Set – Volume 2)

UZUNIDIS Dimitri
Collective Innovation Processes: Principles and Practices
(Innovation in Engineering and Technology Set – Volume 4)

VAN HOOREBEKE Delphine
The Management of Living Beings or Emo-management

2017

AÏT-EL-HADJ Smaïl
The Ongoing Technological System
(Smart Innovation Set – Volume 11)

BAUDRY Marc, DUMONT Béatrice
Patents: Prompting or Restricting Innovation?
(Smart Innovation Set – Volume 12)

BÉRARD Céline, TEYSSIER Christine
Risk Management: Lever for SME Development and Stakeholder Value Creation

CHALENÇON Ludivine
Location Strategies and Value Creation of International Mergers and Acquisitions

CHAUVEL Danièle, BORZILLO Stefano
The Innovative Company: An Ill-defined Object
(Innovation between Risk and Reward Set – Volume 1)

CORSI Patrick
Going Past Limits To Growth

D'ANDRIA Aude, GABARRET Inés
Building 21st Century Entrepreneurship
(Innovation and Technology Set – Volume 2)

DAIDJ Nabyla
Cooperation, Coopetition and Innovation
(Innovation and Technology Set – Volume 3)

FERNEZ-WALCH Sandrine
The Multiple Facets of Innovation Project Management
(Innovation between Risk and Reward Set – Volume 4)

FOREST Joëlle
Creative Rationality and Innovation
(Smart Innovation Set – Volume 14)

GUILHON Bernard
Innovation and Production Ecosystems
(Innovation between Risk and Reward Set – Volume 2)

HAMMOUDI Abdelhakim, DAIDJ Nabyla
Game Theory Approach to Managerial Strategies and Value Creation
(Diverse and Global Perspectives on Value Creation Set – Volume 3)

LALLEMENT Rémi
Intellectual Property and Innovation Protection: New Practices and New Policy Issues
(Innovation between Risk and Reward Set – Volume 3)

LAPERCHE Blandine
Enterprise Knowledge Capital
(Smart Innovation Set – Volume 13)

LEBERT Didier, EL YOUNSI Hafida
International Specialization Dynamics
(Smart Innovation Set – Volume 9)

MAESSCHALCK Marc
Reflexive Governance for Research and Innovative Knowledge
(Responsible Research and Innovation Set – Volume 6)

MASSOTTE Pierre
Ethics in Social Networking and Business 1: Theory, Practice and Current Recommendations
Ethics in Social Networking and Business 2: The Future and Changing Paradigms

MASSOTTE Pierre, CORSI Patrick
Smart Decisions in Complex Systems

MEDINA Mercedes, HERRERO Mónica, URGELLÉS Alicia
Current and Emerging Issues in the Audiovisual Industry
(Diverse and Global Perspectives on Value Creation Set – Volume 1)

MICHAUD Thomas
Innovation, Between Science and Science Fiction
(Smart Innovation Set – Volume 10)

PELLÉ Sophie
Business, Innovation and Responsibility
(Responsible Research and Innovation Set – Volume 7)

SAVIGNAC Emmanuelle
The Gamification of Work: The Use of Games in the Workplace

SUGAHARA Satoshi, DAIDJ Nabyla, USHIO Sumitaka
Value Creation in Management Accounting and Strategic Management: An Integrated Approach
(Diverse and Global Perspectives on Value Creation Set –Volume 2)

UZUNIDIS Dimitri, SAULAIS Pierre
Innovation Engines: Entrepreneurs and Enterprises in a Turbulent World
(Innovation in Engineering and Technology Set – Volume 1)

2016

BARBAROUX Pierre, ATTOUR Amel, SCHENK Eric
Knowledge Management and Innovation
(Smart Innovation Set – Volume 6)

BEN BOUHENI Faten, AMMI Chantal, LEVY Aldo
Banking Governance, Performance And Risk-Taking: Conventional Banks Vs Islamic Banks

BOUTILLIER Sophie, CARRÉ Denis, LEVRATTO Nadine
Entrepreneurial Ecosystems (Smart Innovation Set – Volume 2)

BOUTILLIER Sophie, UZUNIDIS Dimitri
The Entrepreneur (Smart Innovation Set – Volume 8)

BOUVARD Patricia, SUZANNE Hervé
Collective Intelligence Development in Business

GALLAUD Delphine, LAPERCHE Blandine
Circular Economy, Industrial Ecology and Short Supply Chains (Smart Innovation Set – Volume 4)

GUERRIER Claudine
Security and Privacy in the Digital Era (Innovation and Technology Set – Volume 1)

MEGHOUAR Hicham
Corporate Takeover Targets

MONINO Jean-Louis, SEDKAOUI Soraya
Big Data, Open Data and Data Development (Smart Innovation Set – Volume 3)

MOREL Laure, LE ROUX Serge
Fab Labs: Innovative User (Smart Innovation Set – Volume 5)

PICARD Fabienne, TANGUY Corinne
Innovations and Techno-ecological Transition (Smart Innovation Set – Volume 7)

2015

CASADELLA Vanessa, LIU Zeting, DIMITRI Uzunidis
Innovation Capabilities and Economic Development in Open Economies (Smart Innovation Set – Volume 1)

CORSI Patrick, MORIN Dominique
Sequencing Apple's DNA

CORSI Patrick, NEAU Erwan
Innovation Capability Maturity Model

FAIVRE-TAVIGNOT Bénédicte
Social Business and Base of the Pyramid

GODÉ Cécile
Team Coordination in Extreme Environments

MAILLARD Pierre
Competitive Quality and Innovation

MASSOTTE Pierre, CORSI Patrick
Operationalizing Sustainability

MASSOTTE Pierre, CORSI Patrick
Sustainability Calling

2014

DUBÉ Jean, LEGROS Diègo
Spatial Econometrics Using Microdata

LESCA Humbert, LESCA Nicolas
Strategic Decisions and Weak Signals

2013

HABART-CORLOSQUET Marine, JANSSEN Jacques, MANCA Raimondo
VaR Methodology for Non-Gaussian Finance

2012

DAL PONT Jean-Pierre
Process Engineering and Industrial Management

MAILLARD Pierre
Competitive Quality Strategies

POMEROL Jean-Charles
Decision-Making and Action

SZYLAR Christian
UCITS Handbook

2011

LESCA Nicolas
Environmental Scanning and Sustainable Development

LESCA Nicolas, LESCA Humbert
Weak Signals for Strategic Intelligence: Anticipation Tool for Managers

MERCIER-LAURENT Eunika
Innovation Ecosystems

2010

SZYLAR Christian
Risk Management under UCITS III/IV

2009

COHEN Corine
Business Intelligence

ZANINETTI Jean-Marc
Sustainable Development in the USA

2008

CORSI Patrick, DULIEU Mike
The Marketing of Technology Intensive Products and Services

DZEVER Sam, JAUSSAUD Jacques, ANDREOSSO Bernadette
Evolving Corporate Structures and Cultures in Asia: Impact of Globalization

2007

AMMI Chantal
Global Consumer Behavior

2006

BOUGHZALA Imed, ERMINE Jean-Louis
Trends in Enterprise Knowledge Management

CORSI Patrick *et al.*
Innovation Engineering: the Power of Intangible Networks

Printed and bound by CPI Group (UK) Ltd, Croydon, CR0 4YY

13/09/2026

14961035-0007